WELDING AND RELATED TECHNOLOGIES

Welding and Related Technologies contains the papers presented at the VIIth International Conference on "Welding and Related Technologies" (WRT 2024, Yaremche, Ukraine, 7-10 October 2024). The topics covered include:

- Innovative Technologies, Materials and Equipment for Welding and Related Processes
- Strength and Stress-Strain States of Welded Joints and Structures
- Additive Technologies
- New Structural and Functional Materials, Nanomaterials, Composites
- Surface Engineering
- Mathematical Modeling of Welding and Related Processes
- Advanced Technologies of Special Electrometallurgy
- Non-Destructive Testing and Technical Diagnostics
- Materials and Technologies for Medicine and Ecology of Welding Production

Welding and Related Technologies is of interest to scientists from academic institutions, universities and industry, domain experts, and postdoctoral fellows interested in the latest theoretical developments and applications of welding.

PROCEEDINGS OF THE VIIth INTERNATIONAL CONFERENCE ON WELDING AND RELATED TECHNOLOGIES (WRT 2024), 7-10 OCTOBER 2024, YAREMCHE, UKRAINE

Welding and Related Technologies

Edited by

Igor Krivtsun

E.O. Paton Electric Welding Institute of the National Academy of Sciences of Ukraine, Kyiv, Ukraine

Fuad Khoshnaw

De Montfort University, Leicester, UK

Ganna Polishko, Serhiy Schwab & Roman Selin

E.O. Paton Electric Welding Institute of the National Academy of Sciences of Ukraine, Kyiv, Ukraine

CRC Press
Taylor & Francis Group
Boca Raton London New York Leiden

CRC Press is an imprint of the
Taylor & Francis Group, an **informa** business
A BALKEMA BOOK

First published 2025
by CRC Press/Balkema
4 Park Square, Milton Park, Abingdon, Oxon, OX14 4RN

and by CRC Press/Balkema
2385 NW Executive Center Drive, Suite 320, Boca Raton FL 33431

CRC Press/Balkema is an imprint of the Taylor & Francis Group, an informa business

Typeset by Integra Software Services Pvt. Ltd., Pondicherry, India

British Library Cataloguing-in-Publication Data
A catalogue record for this book is available from the British Library

Library of Congress Cataloging-in-Publication Data
A catalog record has been requested for this book

ISBN: 978-1-032-85176-1 (hbk)
ISBN: 978-1-032-85517-2 (pbk)
ISBN: 978-1-003-51851-8 (ebk)
DOI: 10.1201/9781003518518

Welding and Related Technologies – Krivtsun et al. (Eds)
© 2025 The Editor(s), ISBN 978-1-032-85176-1
Open Access: www.taylorfrancis.com, CC BY-NC-ND 4.0 license

Table of contents

Session 3. Innovative technologies, materials and equipment for welding and related processes

Session 4. Mathematical modeling of welding and related processes

Session 8. Strength and stress-strain states of welded joints and structures

Welding and Related Technologies – Krivtsun et al. (Eds)

Preface

This book provides a comprehensive resource on recent advancements in welding and related technologies. By compiling cutting-edge research and foundational principles, it aims to bridge theoretical knowledge with practical applications, offering insights for students, researchers, and industry professionals.

Welding and related technologies encompass an expansive range of processes that are essential for creating high-performance, durable joints in both metallic and non-metallic materials. The importance of welding spans industries as diverse as transportation, construction, aerospace, and medicine, where reliability and precision are crucial. Additionally, adjacent technologies such as additive manufacturing, special electrometallurgy, surfacing, and coating contribute significantly to the enhancement of material properties and structural integrity. In response to growing demands for lighter, stronger, and more environmentally responsible materials, welding technology continues to evolve as a transformative force across multiple sectors.

We would like to extend our heartfelt gratitude to Serhiy Akhonin, Leonid Lobanov, Serhiy Maksymov, Viktor Shapovalov and Volodymyr Korzhyk for their invaluable support and insights. Special thanks are also due to International Institute of Welding (IIW) and the US Civilian Research and Development Foundation (CRDF Global) for their encouragement and contributions that made this book possible. For the solution of organisational problems and technical support, we are grateful to the International Association «Welding» and the NGO «WRTYS». This book is intended for a diverse audience, including students, researchers, and industry professionals involved in materials science, mechanical engineering, metallurgy, and related disciplines. The chapters offer essential knowledge for those developing, testing, and applying welding technologies and serve as a detailed reference for anyone interested in the role of welding in modern technological advancements.

The book is organized into eight main sections, each delving into a specialized area of welding and related technologies:

- Additive Technologies
- Advanced Technologies of Special Electrometallurgy
- Innovative Technologies, Materials, and Equipment for Welding and Related Processes
- Mathematical Modelling of Welding and Related Processes
- Materials and Technologies for Medicine and Environmental Applications
- Non-Destructive Testing and Technical Diagnostics
- New Structural and Functional Materials, Nanomaterials, Composites
- Strength and Stress-Strain States of Welded Joints and Structures

The book begins with chapters that explore recent innovations in additive manufacturing and special metallurgy, providing readers with foundational insights into new high-quality materials and the advanced processes for joining them. Following this, chapters on mathematical modelling cover computational techniques for accurately simulating welding processes—a crucial tool for addressing challenges in heat transfer, deformation, and stress distribution. In the later sections, we examine advanced materials, non-destructive testing methods, and diagnostic techniques to ensure weld quality and integrity. The final chapters focus on new materials, including nanomaterials and composites, that enhance strength and resilience in welded assemblies, meeting the demands of increasingly complex applications.

Welding and Related Technologies – Krivtsun et al. (Eds)
© 2025 The Editor(s), ISBN 978-1-032-85176-1
Open Access: www.taylorfrancis.com, CC BY-NC-ND 4.0 license

Editor Biographies

Igor Krivtsun is a Professor, Director of the E.O. Paton Electric Welding Institute of the National Academy of Sciences of Ukraine. He is a noted scientist in the field of gas discharge physics and theory of welding processes. He made a prominent contribution to development of the above research areas. His basic studies are dedicated to investigation of the physical phenomena occurring in the low-temperature technological plasma (welding arcs, plasma jets, optical and other types of gas discharge), as well as processes of interaction of the electric arc plasma and laser radiation with a material treated under conditions of arc, plasma, laser and hybrid welding, surfacing and coating. His research interest includes hybrid laser-plasma processes of welding and materials treatment. He worked out the theory of interaction of the focused laser beam and arc plasma with the condensed media, which made it possible to describe from the common standpoints the processes of the arc (plasma), laser and combined laser-plasma effects on compact and dispersed materials. Prof. Krivtsun is a full member of the National Academy of Sciences of Ukraine, and a member of the ASM International. He actively cooperates with various R&D institutions of different countries, regularly presents scientific reports at prestigious national and international conferences and participates in accomplishment of a number of international research projects. He is the Editor-in-Chief of the Paton Welding Journal. He has more than 300 scientific publications, including 5 monographs and 18 patents.

Fuad Khoshnaw is a senior academic and researcher at the School of Engineering and Sustainable Development at De Montfort University in the United Kingdom, where he has served as a senior lecturer since joining in 2018. He holds a PhD in Mechanical and Manufacturing Engineering from Loughborough University, UK, which he was awarded in 2010. Dr Khoshnaw is also recognized as a Senior Fellow of the Higher Education Academy in the UK. In 2005, he became a Chevening Scholar, joining the Chevening Technology Enterprise (CTES) program and completing the Innovation and Technology course at London Business School, UK. To date, Dr Khoshnaw has published over 70 papers in fields including welding, corrosion, biomaterials, additive manufacturing, fracture mechanics, and failure analysis, and he has edited six books on welding and corrosion. Dr Khoshnaw is also an active member of several international editorial boards and serves as a reviewer for academic books and journal submissions. His roles include Reviewer for Advanced Engineering Materials (Wiley), Reviewer for Anti-Corrosion Methods and Materials (Emerald), Reviewer for the Journal of Materials Engineering and Performance (Springer Science, ASM International), Editorial Board Member of the International Journal of Energy Production and Management (WIT Press, UK), Reviewer and Editorial Board Member for the Independent Journal of Management and Production (Brazil), and others.

Ganna Polishko is an Associate Professor, Doctor of Engineering Sciences, Head of Department at the E.O. Paton Electric Welding Institute of the National Academy of Sciences of Ukraine. She is a scientist in the field of processes of special electrometallurgy, electroslag metallurgy, secondary metallurgy, composite and nanomaterials and additive technologies. She studies cleanliness and properties of steels and alloys, behaviour of non-metallic inclusions and metallographic investigations. Her research interests include evaluation of weldability of high alloy steels, advanced high-strength steels, railway steels, power steels and titanium alloys. Dr. Polishko held the position of Associate Professor at the National Technical University of Ukraine "Kyiv Polytechnic Institute" and was an invited lecturer at the University of Lorraine in France. She was a co-editor of the monograph based on the results of the VIII International Conference of Young Scientists on Welding and Related Technologies (WRTYS 2015) and the Proceedings of the IX International Conference of Young Scientists on Welding and Related Technologies (WRTYS 2017). She has served as the head and member of the organising committee for a number of international conferences, including WRTYS 2013, WRTYS 2015, WRTYS 2017, ADMET 2016, ADMET 2018 and WRT 2024. She has also authored over 90 scientific publications and holds two patents.

Serhiy Schwab is a Senior Researcher, Department of Metallurgy and Welding of Titanium Alloys, E.O. Paton Electric Welding Institute of the NAS of Ukraine, Kyiv, Ukraine. He is a specialist in the development of materials and technologies for welding, surfacing and additive manufacturing of titanium and its alloys. He earned a PhD at the E.O. Paton Electric Welding Institute of the NAS of Ukraine in 2018. He earned a master's at the National Technical University of Ukraine "Kyiv Polytechnic Institute" in 2009. He was a lecturer in the Kyiv Mechanical and Technological College (Ukraine) in 2014-2016. Dr. Schwab has received numerous awards from the NAS of Ukraine and from the mayor of Kyiv. He was a co-editor of the monograph is based on the results of VIII International conference of Young Scientist «Welding and Related Technologies» (WRTYS 2015) and the Proceedings of the IX International conference of Young Scientists on Welding and Related Technologies (WRTYS 2017). He has published 20+ articles. He has participated with scientific presentations in important scientific conferences, including those under the auspices of the International Institute of Welding (annual Assemblies). For special merits he received scholarships from the President of Ukraine and from the NAS of Ukraine for young scientists. He is the organizer of a number of conferences: WRTYS 2017, WRT 2018, YPIC and WRTYS 2021 and WRT 2024. He received research grants from the NAS of Ukraine, CETS (Hungary) and French Ambassy in Ukraine.

Roman Selin is a Senior Researcher at the E.O. Paton Electric Welding Institute of the National Academy of Sciences of Ukraine, joining the institute in 2011. He holds a PhD in Material Science, which was awarded in 2016. Currently, his research focuses on welding, additive manufacturing, mathematical modelling, mechanical and structural properties of titanium alloys. Dr. Selin has published extensively on these topics, sharing his findings with the scientific community and furthering innovations in welding science, which resulted in over 50 papers. He also was a reviewer for the *Proceedings of IX International Conference of young scientists "Welding and related technologies. WRTYS-2017" (Ukraine).*

Session 1. Additive technologies

Welding and Related Technologies – Krivtsun et al. (Eds)

Study of the influence of manufacturing parameters on the mechanical properties of the LPBF-AlSi10Mg alloy

S.V. Adzhamskyi
Institute of Transport Systems and Technologies of the National Academy of Sciences of Ukraine, Dnipro, Ukraine
Additive Laser Technology of Ukraine, Odessa, Ukraine

G.A. Kononenko & R.V. Podolskyi
Iron and Steel Institute of Z.I. Nekrasov of National Academy of Sciences of Ukraine, Dnipro, Ukraine
Additive Laser Technology of Ukraine, Odessa, Ukraine

ABSTRACT: The work is devoted to the study of the influence of the circulation of the protective medium in the working space of the printer camera during laser powder layer melting (LPBF). According to the results of research, the influence of the speed and trajectory of blowing is especially relevant for materials with low specific density. The material for research is LPBF-AlSi10Mg alloy samples, evenly spaced along the working plane of the camera, manufactured according to rational technological modes. Tensile tests were carried out according to standard methods. Continuity was determined metallographically using ImageJ software. It was established that when supplying a constant feed rate and circulation of the protective medium, alloy particles are captured and transferred to the exit nozzle, which affects the final integrity, namely the pore size. The dependence of mechanical properties on the location of the studied samples in the platform area is shown.

1 INTRODUCTION

Additive manufacturing technology, also known as 3D printing, has recently been increasingly used, and the number of materials and methods that can be used in various industries such as aerospace, automotive, medical, etc. is expanding.

In this work, we focus on laser powder bed fusion (LPBF). LPBF is an iterative process consisting of three main steps: (1) depositing a 15- to 150-μm powder layer on the build platform; (2) partial melting of the powder layer by a laser source based on previously imported 3D-CAD data; and (3) lowering the construction platform and restarting at (1). The powder is usually applied with a polymer or rubber scraper. The manufacturing process is carried out in a circulating protective environment in order to protect against chemical reactions and the formation of by-products of the process in the working space of the camera.

Research in recent years has mainly focused on process control, including the influence of various process parameters on the stability and resulting microstructure and material properties (Liverani E. et al. 2017, Yadroitsev I. et al. 2013, Yadollahi A. et al. 2015). In recent years, much attention has been paid to the speed and circulation of the protective medium and their interaction with various materials of different density and, in turn, weight. From the results of the authors' work (Anwar A. & Pham Q.-C. 2017, Avegnon K. L. M. et al. 2023), it was established that the speed and circulation play a significant role in the integrity and final mechanical properties of the test samples located along the cross-section of the platform at equal distances.

DOI: 10.1201/9781003518518-1

For the production of parts, rational technological parameters are often used for the corresponding device and material, as provided by the manufacturer. Various properties of materials for samples made using LPBF have been determined in the course of many laboratory studies (Qiu C. et al. 2015, Yang Y. et al. 2020). These works often used their own optimized process parameters, so these values do not necessarily reflect achievable values in an industrial process.

These studies provide an understanding of the influence of many factors present in additive manufacturing processes using a powder layer, such as powder layer density and flow behavior or heat flow and temperature distribution (Haeri S. et al. 2017, Du Y. et al. 2019). In addition, simulation studies of defect formation and microstructure during these processes were conducted (Ning J. et al. 2019, Ning J. et al. 2020, Zhang Y. & Zhang J. 2019).

2 METHODOLOGY

In the section, the main laboratory-methodical processes of manufacturing test samples from LPBF-AlSi10Mg aluminum-based alloy, equipment and testing methodology for determining the mechanical properties and integrity of the test samples were considered.

2.1 *Production of experimental samples from LPBF-AlSi10Mg alloy*

The production of experimental samples was carried out on the Alfa-150D production plant of the company "ALT Ukraine" LLC, which is equipped with an F-Theta lens and a system of combined blowing with a protective gas (argon) with a circulation speed of 1.5 m/s, as a source – ytterbium laser with a wavelength of 1064 nm, construction platform - 150x150 mm. The material used in this study was an aluminum-based alloy LPBF-AlSi10Mg with a particle size from 10 to 45 μm, which allows determining the blowing conditions for integrity and mechanical properties of light metals (with a low specific density). Chemical composition of AlSi10Mg powder in % by weight: Si=10.0; Mg=0.35; Mn=0.43; Al-balance. The starting material was examined using a scanning electron microscope REM-106 (Figure 1a) to determine the shape and size of particles. Ha Figure 1b shows the results of granulometric analysis.

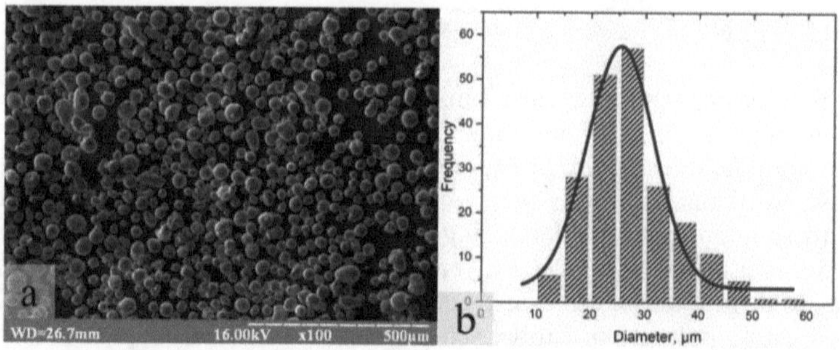

Figure 1. Particles of the starting material AlSi10Mg at a magnification of 100 (a) and the results of granulometric analysis (b).

Samples 10×10×5 mm in size, manufactured using a constant layer thickness of 40 μm and variable scanning parameters (power, scanning speed, distance between tracks), namely 180 and 195 W; 1200…1300 mm/s with a step of 50 mm/s; 0.11…0.13 mm in steps of 0.01 mm, respectively. Samples for determining the characteristics during tensile tests were made according to the parameters: power - 180 W, scanning speed - 1200 mm/s and the distance between the tracks 0.11 mm, which were determined as rational from the results of the continuity analysis. The tensile test specimens were placed vertically on the working platform at equal distances from each other. Thus, allowing to analyze the effect of the change in properties depending on the location on the platform and the interaction of the effect of the circulation of the protective medium on the manufacturing process.

2.2 *Microstructural characterisation*

Metallographic studies of steels were carried out with the help of light microscopes manufactured by Carl Zeiss company, Neophot 32 and Axiovert 200 M MAT models. The preparation of the grindings was carried out mechanically using specialized equipment for the preparation of metallographic grindings of the company "Struers" model "Labopol-5" with the use of the necessary diamond suspensions. The porosity of the samples under investigation was characterized by analyzing the polished surfaces using the ImageJ software and using the "Particle Analysis" function. Structure objects smaller than 4 μm were filtered out, excluding these elements as sample preparation defects. Identification of the microstructure was carried out using organic benzene mixtures. Electrochemical polishing of samples was performed using oxalic acid ($C_2H_2O_4$).

2.3 *Mechanical properties*

Tensile tests were performed on cylindrical samples with a working zone diameter of 5 mm, manufactured in accordance with DSTU ISO 6892-1:2019. The tests were carried out on a specialized Instron testing machine with a maximum load of 1 t and a traverse speed of 0.01 mm/s with fixation of the engineering curve in stress/strain coordinates.

3 RESULTS

From the results of studies (Figure 2) of mechanical properties (Figure 2a, c) and integrity analysis (Figure 2 d) depending on the location on the platform (Figure 2 d) it was established that all test samples except for sample 3 and 5, which were located near the exit nozzle, have approximately the same density over the total area of the experimental area. It should be noted that the discontinuities in test samples 3 and 5 have a smaller number, but the pore size is in the range of 3…12 μm, while the other test samples are 3…8 μm with a larger total number. Thus, the mechanical properties showed that, taking into account the discontinuities in these samples 3 and 5, the overall

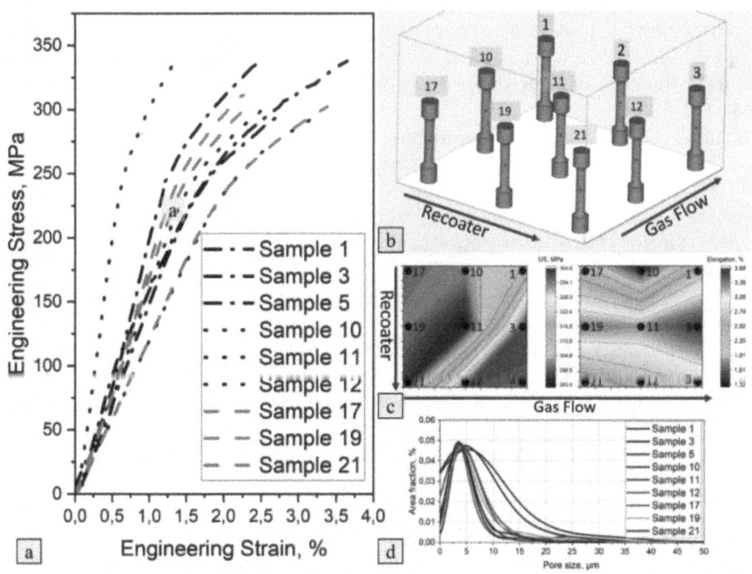

Figure 2. The stress/strain diagram (a), the layout of the test samples on the platform (b), the change in mechanical properties (c) and the relative total proportion of pores (d) in the test samples depending on the location on the platform.

porosity plays a more significant role than the pore size, which explains the high strength (340...334 MPa) and medium plasticity (3.09...3.68%) characteristics. It should be noted that the test samples 10 - 11, 17, 18, 21 mainly have strength - 293...304 MPa, the plasticity of the test samples corresponds to the range of 2.79...3.09%. Strength and plasticity characteristics of test sample 10 below, as is known from work (Adzhamskyi S.V. et al. 2023), this is due to the circulation of the protective medium in this area.

Thus, as a result of the research, it was established that when applying a constant feed rate and circulation of the protective medium, they affect the grains of an alloy with a low density, which, accordingly, has a lower weight, namely the LPBF-AlSi10Mg alloy. It was shown that the capture and transfer of particles of this alloy along the path of circulation of the protective medium does not have a significant effect on the mechanical properties, which was also proven in works [4-5]. It should be noted that the particle transfer effect affects the density of the test samples located in the area of the outlet nozzle, namely the pore size.

4 CONCLUSIONS

1. It was established that when applying a constant feed rate and circulation of the protective medium, particles are picked up and transferred to the exit nozzle, which affects the final integrity, namely the pore size, of the LPBF-AlSi10Mg alloy.
2. The dependence of the change in mechanical properties on the location of the test samples in the area of the platform and the influence of their blowing are shown. As a result, it was established that mainly the strength of all test samples corresponds to the range of values of 340...293 MPa, and plasticity - 2.2...3.68%.

REFERENCES

Liverani E. et al. 2017. Effect of selective laser melting (SLM) process parameters on microstructure and mechanical properties of 316L austenitic stainless. *Journal of Materials Processing Technology* 249: 255–263.

Yadroitsev I. et al. 2013. Energy input effect on morphology and microstructure of selective laser melting single track from metallic powder. *Journal of Materials Processing Technology* 213: 606–613.

Yadollahi A. et al. 2015. Effects of process time interval and heat treatment on the mechanical and microstructural properties of direct laser deposited 316L stainless steel, *Materials Science and Engineering: A* 644:171–183.

Anwar A. & Pham Q.-C. 2017. Selective laser melting of AlSi10Mg: Effects of scan direction, part placement and inert gas flow velocity on tensile strength. *Journal of Materials Processing Technology* 240: 388–396.

Avegnon K. L. M. et al. 2023. Porosity formation from disrupted gas flow in laser powder bed fusion of 316 stainless steel. *Journal of Manufacturing Processes* 90: 333–340.

Qiu C. et al. 2015. On the role of melt flow into the surface structure and porosity development during selective laser melting. *Acta Materialia* 96: 72–79.

Yang Y. et al. 2020. A computationally efficient thermal model for selective laser melting. *Additive Manufacturing* 31: 100955.

Haeri S. et al. 2017. Discrete element simulation and experimental study of powder spreading process in additive manufacturing. *Powder Technology* 306: 45–54.

Du Y. et al. 2019. A model for predicting the temperature field during selective laser melting, *Results in Physics* 12: 52–60.

Ning J. et al. 2019. Analytical modeling of lack-of-fusion porosity in metal additive manufacturing. *Applied Physics A* 125: 1–11.

Ning J. et al. 2020. Analytical modeling of part porosity in metal additive manufacturing. *International Journal of Mechanical Sciences* 172: 105428.

Zhang Y. & Zhang J. 2019. Modeling of solidification microstructure evolution in laser powder bed fusion fabricated 316L stainless steel using combined computational fluid dynamics and cellular automata. *Additive Manufacturing* 28: 750–765.

Adzhamskyi S.V. et al. 2023. Influence of blowing and loading of working space on mechanical properties of samples manufactured using SLM technology. *The Paton Welding Journal*, 12: 19–23.

Welding and Related Technologies – Krivtsun et al. (Eds)

Application of additive electron-beam technologies for aviation and medical needs

V.A. Matviichuk & V.M. Nesterenkov
E.O. Paton Electric Welding Institute, Ukraine

ABSTRACT: Innovative technologies of layer-by-layer production of products by the method of rapid prototyping open up new opportunities for the production of parts of a given shape and structure with properties predicted in advance. The processes of manufacturing products by this method using an electron beam have shown great prospects for their application for the production of a wide range of parts and assemblies. On the basis of industrial equipment for electron beam welding, experimental equipment for reproduction of additive processes has been created, and a hardware and software platform for controlling additive production has been developed. Invented technological modes and methods of printing products. According to the additive technology, prototypes were manufactured and tested, and rational technological regimes of surfacing were invented. Methods of designing products have been created. Printing of experimental samples for the needs of aircraft engine construction was carried out. Designed and manufactured individual medical implants.

1 INTRODUCTION

Innovative technologies of layer-by-layer manufacturing using rapid prototyping methods open new possibilities for producing parts with predetermined shapes and structures and forecasted properties. Manufacturing processes using electron beams have shown great potential for producing a wide range of parts and assemblies (Nesterenkov et al. 2019).

The development of equipment and software for additive electron-beam manufacturing, particularly for implementation in aerospace industry enterprises, is highly relevant (Matviichuk & Nesterenkov 2020, Matviichuk et al. 2022).

In the context of martial law in Ukraine, it is crucial to provide the medical sector with reliable and competitive domestic implants and prosthetics. Additive methods are the most effective for producing customized medical devices.

These issues highlight the need for scientific research on the application of additive electron-beam technologies for aviation and medical purposes.

The aim of the work is to develop additive equipment and technology for manufacturing products using layer-by-layer electron-beam deposition with metallic powder materials.

To achieve this aim, the following tasks were set:

- to identify the stages of additive electron-beam manufacturing;
- to develop experimental equipment;
- to create a software and hardware platform for managing additive manufacturing;
- to investigate the impact of technological parameters of deposition on the properties of the products;
- to develop methods for designing and printing products;
- to create an additive electron-beam technology for layer-by-layer manufacturing of products from metallic powder materials.

The sequence of stages in additive electron-beam manufacturing is shown in Figure 1 (Matviichuk et al. 2022).

DOI: 10.1201/9781003518518-2

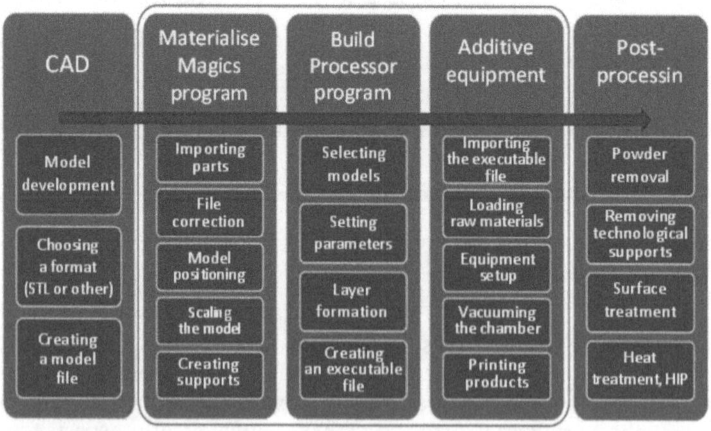

Figure 1. Stages in additive electron-beam manufacturing.

A 3D model of the object is created for printing, which is converted into a standard format for additive manufacturing, usually an STL file. Next, a product assembly is created, which is divided into layers using software tools.

The next stage requires transferring the file to the machine and configuring it.

Then the machine, controlled by a computer, builds the products layer by layer. The size of the products depends on the size of the machine.

After construction and cooling, the assembly can be removed. Then, the cleaning and surface finishing of the parts are performed. Additional processing of the products to meet the required standard is also possible. This requires the use of other machines and tools.

To conduct research based on compact equipment for electron beam welding of the CV-212M type, experimental additive equipment was created (Figure 2) (Matviichuk & Nesterenkov 2020).

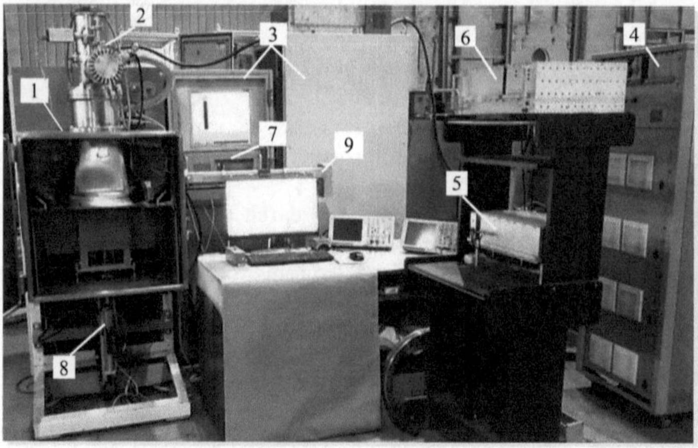

Figure 2. Composition of the additive equipment: 1 – vacuum chamber, 2 – electron gun, 3 – cabinets, 4 – high-voltage source, 5 – sweep amplifier, 6 – dynamic focusing amplifier, 7 – industrial computer, 8 – tail spindle, 9 – actuator.

The machine consists of a compact vacuum chamber (1) with platform movement mechanisms (8), as well as powder feeding and distribution mechanisms (9). The power supply unit includes an electron beam gun and a high-voltage power source. The electron beam gun (2) is located at the top of the vacuum chamber. The equipment's vacuum system ensures a vacuum in the chamber up to 0.01 Pa. The control system components are located in the units (3). The high-voltage source (4) allows adjustable voltage up to 60 kV and electron beam current up to 100 mA. The sweep signal amplifier (5) and dynamic focusing amplifier (6) shape the electron beam sweeps and its focusing. The equipment is controlled by an industrial computer (7).

A hardware-software platform for controlling additive manufacturing was developed in collaboration with Materialise, Belgium, consisting of a control unit and an application software package (Figure 3) (Matviichuk & Nesterenkov 2020).

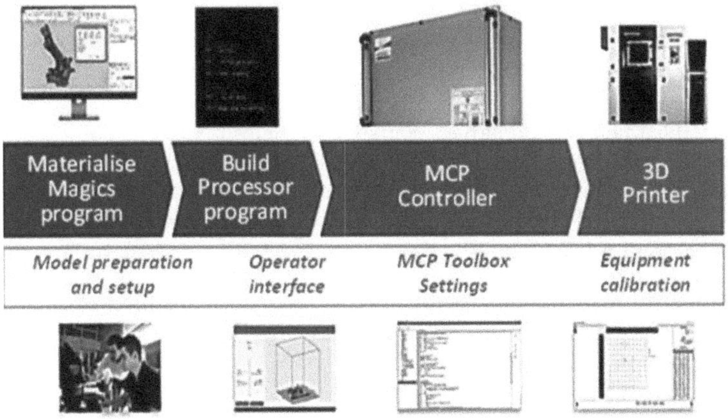

Figure 3. Structure of the control platform.

A concept for the hardware-software platform has been developed. A control unit with an application software package for 3D printing was designed, manufactured, and installed on the additive equipment.

The program Materialise Magics is a universal software for preparing data for additive manufacturing, which serves to analyze and edit 3D models, check and correct geometry errors, and create product assemblies and structuring.

The Build Processor program allows the models to be sliced into layers, set the parameters and structure of their structuring, and determine the power and speed of the electron beam movement for each product, as well as focusing parameters and layer texture filling options.

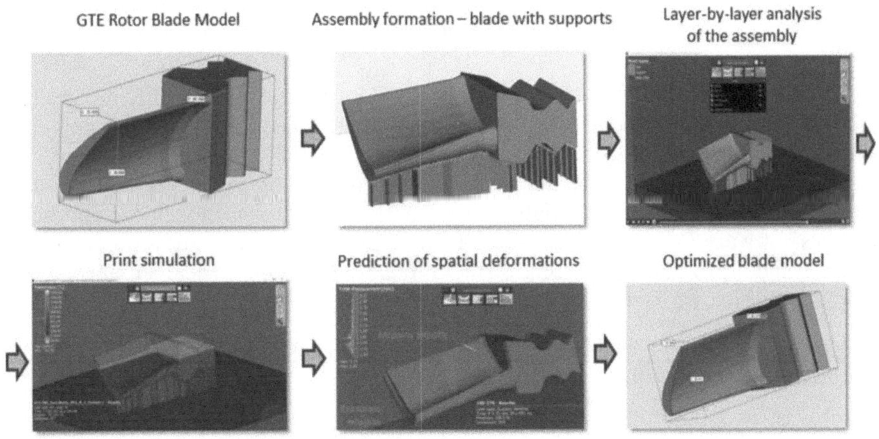

Figure 4. Process of creating a turbine blade model for a gas turbine engine (GTE).

1.1 *Design of aircraft engine blade models*

Digital models of the rotor blade of an aircraft engine were designed in the SOLIDWORKS Desktop 3D CAD program (Solidworks 2024) (Figure 4). In the Materialise Magics program (Materialise Magics 2024), an assembly was created, including the product with technological supports. In the Simufact Additive program (Derrer et al.), a layer-by-layer analysis of the assembly and simulation of the printing processes were performed, the influence of technological parameters on the properties of the blade was studied, the product's shape formation during printing was predicted, and optimized blade models were obtained (Sahini et al. 2020). The corrected model is exported in STL format and prepared for printing.

1.2 *Development of methods for designing and manufacturing individual medical implants using additive methods*

Ensuring the medical field with reliable, competitive domestic implants and prosthetics is crucial and highly relevant in the context of military operations in Ukraine. To address this issue, methods for designing individual implants have been developed, and samples of medical products have been printed. The work was carried out in collaboration with the Irpin Military Hospital and Materialise Ukraine.

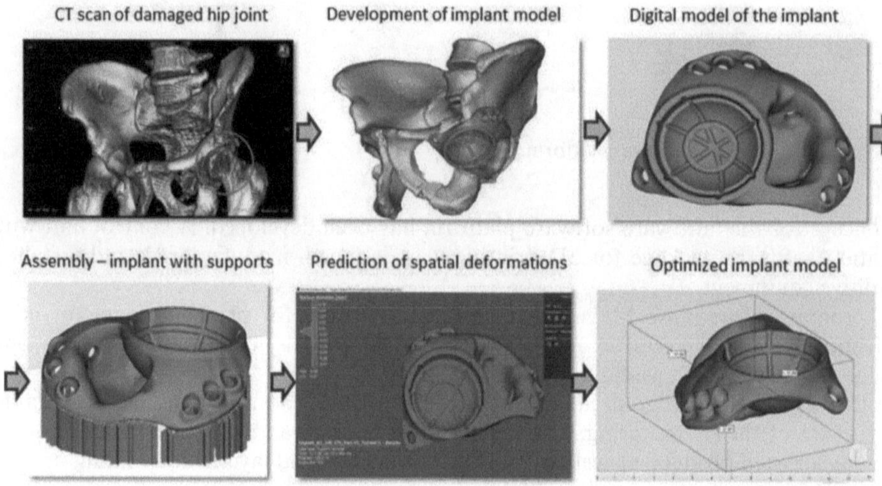

Figure 5. Process of creating a medical implant model.

Based on the results of a computed tomography (CT) scan of a patient with impaired walking function, the tomogram of the affected organ was analyzed in the Radiant program. The joint was damaged as a result of combat actions, and the spatial positioning of the bones was changed. Treatment involves restoring walking function through prosthetics using a custom implant. In collaboration with Materialise Ukraine, the implant was modeled using the Materialise Mimics software (Materialise Mimics 2024) (Figure 5). An assembly was created in the Materialise Magics program, consisting of the implant with technological supports. Optimized implant models were obtained using the Simufact Additive program.

1.3 *3D printing and evaluation of aircraft parts*

An assembly of two GTE blades and four calibration samples was created using Materialise Magics program (Figure 6a).

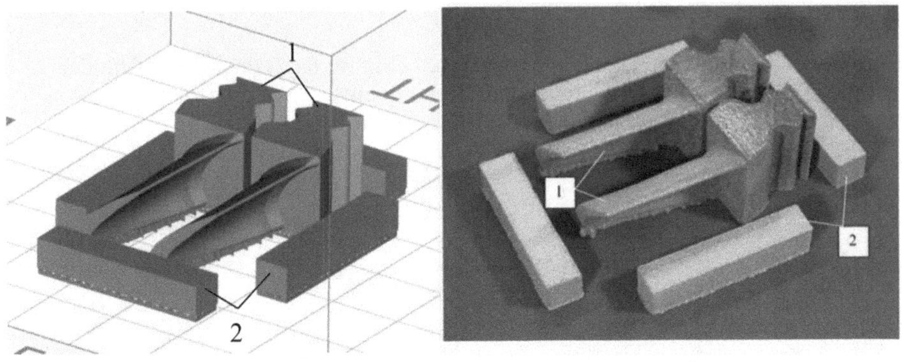

a b

Figure 6. Computer model (a) and printed products (b) from titanium alloy TA15: 1 – GTE rotor blade, 2 – test samples.

Using titanium alloy TA15 powder and under specified technological regimes, the products were printed (Figure 6b). Optimized models were used for the printing process.

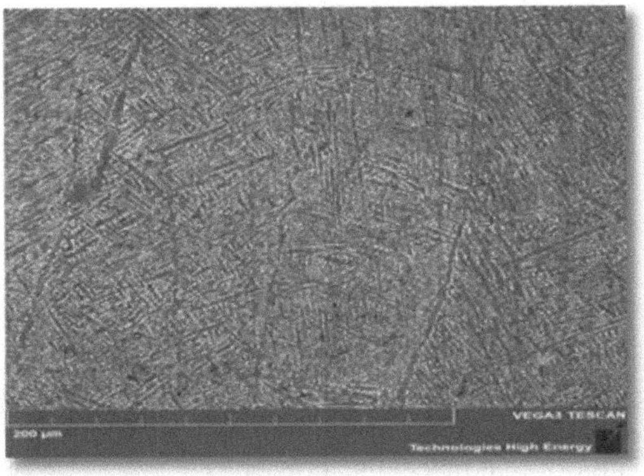

Figure 7. Microstructure of the test sample made from TA15 titanium alloy.

The test sample was examined. It was determined that the alloy's structure is thin-lamellar and biphasic (Figure 7), which is typical for the TA15 titanium alloy. The chemical composition corresponds to TA15. The Rockwell hardness is around 40 HRC. The sample's porosity does not exceed 0.45%.

Additionally, samples produced using electron beam 3D printing (EBP) (Matviichuk et al. 2022) and samples manufactured using electron beam melting (EBM) technology (Akhonin et al. 2022) were studied. The strength characteristics of the TA15 alloy produced by EBP technology are 20% higher than those of cast metal produced by EBM technology. The plasticity values when using EBP technology are three times higher compared to EBM technology. This is due to the formation of a finer microstructure in the products obtained through 3D printing technology (Akhonin et al. 2024).

The impact toughness value is KCU 55 J/cm² for samples made from titanium alloy ingots using EBM technology. This is almost twice as high as KCU 33 J/cm² for samples created by EBP. The decrease in impact bending test results occurs due to unfavorable orientation of the sample relative to the direction of applied load, caused by the structuring of a unidirectional dendritic metal structure (Akhonin et al. 2024).

1.4 *Manufacturing of custom medical implants*

An assembly of products was created for research (Figure 8a), consisting of models of a hip joint implant, a maxillofacial implant, biomedical implants, and a control sample.

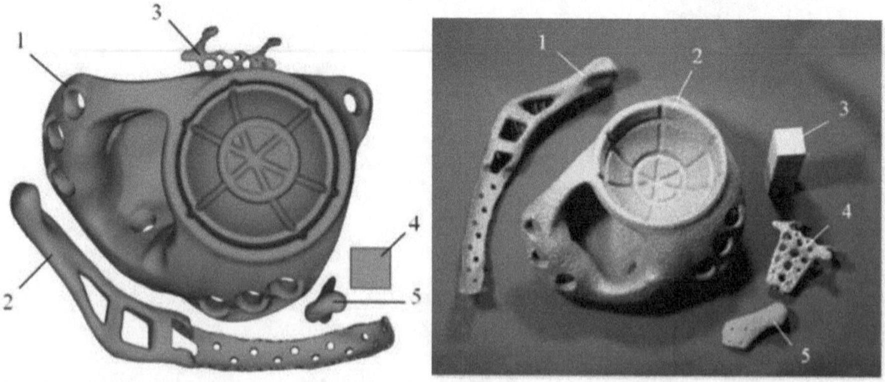

Figure 8. Computer model (a) and printed products (b) from TA15 titanium alloy: 1 – hip joint implant, 2 – maxillofacial implant, 3, 5 – biomedical implants, 4 – control test sample.

Medical products – custom implants – were manufactured from the TA15 titanium alloy using the additive electron beam deposition method according to the specified technological parameters (Figure 8b).

2 CONCLUSIONS

1. Additive equipment has been created for manufacturing metal products layer by layer based on a digital computer model.
2. A hardware-software platform for managing additive manufacturing has been developed.
3. Printing regimes for high-strength TA15 titanium alloys have been identified, ensuring the best metal structure and optimal mechanical properties.
4. Methods for designing products for aviation and medical needs have been developed, with prototypes of GTE blades and custom medical implants printed, and the properties of control samples investigated.

REFERENCES

Akhonin, S., Pikulin, O., Berezos, V., Severyn, A., Erokhin, O., Kryzhanovskyi, V. 2022. Determining the structure and properties of heat-resistant titanium alloys VT3-1 and VT9 obtained by electron-beam melting. *Eastern-European Journal of Enterprise Technologies* 5 (12 (119)): 6–12.

Akhonin, S., Nesterenkov, V., Pashynskyi, V., Matviichuk, V., Motrunich, S., Berezos, V., Klochkov, I. 2024. Determining technological parameters for obtaining ta15 titanium alloy blanks with improved mechanical characteristics using the electron-beam 3D printing method. *Eastern-European Journal of Enterprise Technologies* 3(12 (129)): 36–45.

Derrer, J.P., Escobar de Obaldia, E., Mehmert, P., Schafstall, H. 2020. Simufact Engineering provides the GreenTeam & Renishaw with a complete AM Process Simulation Solution. *SIMUFACT Case study*.

Materialise Magics 2024 (Electronic source). https://www.materialise.com/en/industrial/software/magics-data-build-preparation

Materialise Mimics 2024 (Electronic source). https://www.materialise.com/en/healthcare/mimics-innovation-suite/mimics

Matviichuk, V.A. & Nesterenkov, V.M. 2020. Additive electron beam equipment for layer-by-layer manufacture of metal products from powder materials. *The Paton Welding Journal* 2: 41–46.

Matviichuk, V. A., Nesterenkov, V. M., Berdnikova, O. M. (2022). Additive electron beam technology of manufacture of metal products from powder materials. *The Paton welding journal* 2: 16–25.
Nesterenkov, V.M., Matviichuk, V.A., Rusynik, M.O., Yanko, T.B., Dmitrenko, A.E. 2019. Microstructure of VT20 titanium alloys produced by the method of layer-by-layer electron beam fusion using domestic powder materials. *The Paton welding journal* 9: 2–7.
Sahini, D., Ghose, K., J., Jha, S. K., Behera, A., Mandal A. 2020. Optimization and Simulation of Additive Manufacturing Processes: Challenges and Opportunities – A Review. In book: *Additive Manufacturing Applications for Metals and Composites*: 187–209. Hershey PA: IGI Global.
SOLIDWORKS 3D CAD 2024 (Electronic source). https://www.solidworks.com/product/solidworks-3d-cad

Welding and Related Technologies – Krivtsun et al. (Eds)

The effect of heat treatment on Ti-Zr-Nb alloy multilayer joints, obtained by the WAAM with filler cored wire

S.L. Schwab, R.V. Selin & A.A. Babinets
E.O. Paton Electric Welding Institute of the NAS of Ukraine, Kyiv, Ukraine

M.M. Voron
Physical and Technological Institute of Metals and Alloys of the NAS of Ukraine, Kyiv, Ukraine

T.B. Yanko
PJSC Titanium Institute/Underslab LLC, Burgas, Bulgaria

ABSTRACT: The multilayer sample based on titanium alloy of the Ti-Zr-Nb system was obtained by the WAAM (TIG method) with filler metal-powder wire, the basis of which is powder obtained by the Hydrogenation-DeHydrogenation method. As a result, a 9-layer sample was produced. Its microhardness is at the level of 4200-5000 MPa. The effect of various types of heat treatment (vacuum annealing, quenching, quenching with aging) on the microstructure and microhardness of the obtained metal was studied.

1 INTRODUCTION

For all structural materials, there is always a need to constantly improve their reliability and durability. Such a specific variety of materials as alloys for endoprosthesis must meet a greater number of requirements, compared to more common alloys of the wide use. First of all, they must meet high requirements for biocompatibility, so they must not contain toxic components. Another important condition is an optimal set of mechanical properties, among which the main attention is paid to high strength and low modulus of elasticity. In an optimal way, such conditions are most mostly fulfilled when using titanium alloys (Wang 1996).

A large number of researchers from all over the world separate only a small amount of biocompatible metals, besides titanium, such as Zr, Nb, Sn, Ta and Mo (Silva et al. 2004).

Titanium alloys of the Ti-Zr-Nb and Ti-Zr-Nb-Sn systems are most reliable from the point biocompatibility and complex mechanical properties. For example, Ti-13Nb-13Zr, Ti-24Nb-12Zr and Ti-(24-26)Nb-(4-8)Zr-(7-8Sn) alloys are increasingly used in endoprosthesis. Their strength limit is at the level of 800-1000 MPa, and the modulus of elasticity is at the level of 45-65 GPa (Niinomi 2008, Yoshimitsu et al. 1993, Shneider et al. 2005).

Having analyzed the relevance of using low-modulus titanium alloys for endoprosthesis and the technology of manufacturing finished products from them, which involve the presence of rods for fixation, or other loaded and massive elements, a conclusion was made about the feasibility of the Wire Arc Additive Manufacturing (WAAM) method (Sokolovskyi & Bernatskyi 2023) as an alternative to existing additive manufacturing methods.

DOI: 10.1201/9781003518518-3

2 MATERIALS AND METHODS

2.1 *Preparation of materials and production of cored wire based on Ti-Zr-Nb system alloy*

The most versatile and multifunctional casting method for obtaining castings from titanium and its alloys is the electron-beam casting technology (electron-beam garnish melting) (Ladokhin et al. 2020). This method allows for the preparation of the wide range of alloy melts based on refractory and highly reactive metals, alloys based on intermetallics, special and complex titanium alloys, which include the biomedical alloy of Ti-Zr-Nb systems. According to the results of melting, a high-quality ingot of Ti-13Zr-13Nb alloy with a diameter of 55 mm and a length of 250 mm was obtained. Local EDX analysis of the sample showed a uniform distribution of elements.

One of the important and complex technological processes in the production of metal powder wires is the preparation of their core - the metal component, that is, the powder itself. The powder from this alloy was produced by the Hydrogenation-DeHydrogenation (HDH) method (Yanko et al. 2018). To obtain Ti-13Zr-13Nb alloy powder, the method of hydrogenation of the cast alloy was used. In the process of hydrogenation, the workpiece was heated to a temperature of about 900 °C, followed by hydrogen treatment. In the process of hydrogen saturation, deformation of crystal lattices took place, which led to embrittlement of the material. Due to this development, it was possible to grind the alloy into powder. After grinding the alloy, the powder was classified into fractions. The fraction (5...50 μm) required for the research was subjected to thermal vacuum treatment to remove hydrogen from the powder.

The development and production of metal cored wire based on Ti-13Zr-13Nb titanium alloy was carried out by the metal dragging method on the installation, which is designed for the production of titanium cored wires (Schwab et al. 2023, Akhonin & Schwab 2019, Ryabtsev et al. 2023). VT1-00 titanium foil with a thickness of 0.2 mm was used as the shell. Pulling was carried out using a set of dies with a step of 0.1 mm. As the result, the wire with 3.0 mm diameter was obtained, with a filling factor of 65%.

Multilayer TIG surfacing with the obtained metal cored wire based on biomedical titanium alloy was carried out on the equipment designed for TIG welding and surfacing of titanium-based alloys (Figure 9). The process was carried out on a pre-determined mode: Iw, = 210 A, Uarc = 12,7 V, Vweld = 8 m/h, Vwire = 34 m/h, Larc = 3,5 mm.

2.2 *Methods used in research*

The content of oxygen, nitrogen and hydrogen was determined by the method of reduction melting in the flow of the carrier gas. The oxygen content was determined on LECO RO-316; nitrogen content was determined on LECO TN-114; hydrogen content was determined on LECO RH-2.

Microstructure studies were performed by electron microscopy on Tescan Vega 3 scanning electron microscope. It allows to obtain a topographic image in the mode of detection of secondary electrons (SE - secondary electrons) and an image by contrast according to the atomic weight of the components in the mode of detection of reflected electrons (BSE - backscattered electrons). Preference was given to the second mode. During microscopic studies, local chemical analysis of phases and EDX-mapping (color elemental distribution by their atomic weight) were determined by Bruker detector, installed on microscope.

Studies of microhardness were carried out on a microhardness tester PMT-3 with a load on the indenter of 100 g.

3 RESULTS AND DISCUSSION

As a result of the experiments a 9-layer sample was obtained with metal cored wire based on the powder of the Ti-13Zr-13Nb alloy. The wire melting process for this case was stable, the layers were deposited with a constant height (on average, the height of each layer is

1.6-1.8 mm), the width of the sample is about 11 mm and 15.2 mm high. The base is 6 mm thick pure titanium.

The X-ray deffectoscopy did not reveal any significant defects, except for a small sized pores (up to 50 μm), which is allowed in titanium welded joints. They are mostly located in the middle part of each layer.

Gas analysis was carried out for the ingot itself, for the powder and for the surfaced metal (Table 1). A significant increase in oxygen in the material is undoubtedly associated with a significant increase in the specific area of the material. Each particle of the alloy powder was covered with an oxide layer since titanium and zirconium are very prone to oxidation. It is possible that during the hydrogenation process, local overheating occurred during the exothermic hydrogenation reactions of titanium and zirconium. Some recrystallization of the alloy may have also occurred during grinding. This could've lead to local phase transitions with the formation of structures based on the Ti-Zr-Nb system, which have an increased affinity for oxygen. The decrease in oxygen content after the surfacing process is due to the fact that pure titanium foil was added to the content of the composite wire.

Table 1. Results of gas analysis of samples.

Research area	[O] / wt%	[N] / wt%	[H] / wt%
Casting of the Ti-13Zr-13Nb alloy	0,82	0,034	0,026
HDH powder of Ti-13Zr-13Nb alloy	2,76	0,61	0,0157
The metal of the surfaced sample	1,35	0,072	0,0063

Studies of the elements distribution by the EDX-mapping method showed high uniformity of the deposited metal. Analysis of the chemical composition of the middle part of the sample shows the zirconium and niobium content of 6.53 and 6.67% by weight accordingly. The decrease in the number of these elements is due to the additional alloying with the pure titanium shell of the wire.

Subsequently, the received sample was subjected to various types of heat treatment:

1) Vacuum annealing (heating to 750 °C, holding for 1 hour, cooling in air).
2) Quenching (heating to 960 °C, holding for 1 hour, cooling in water).
3) Quenching with aging (heating to 960 °C, holding for 1 hour, cooling in water, heating to 450 °C, holding for 4 hours, cooling in water).

The microstructure study of the sample's central part in the state "after surfacing" (Figure 1a) showed that it is characterized by relatively small sizes of primary β-grains, but also by the simultaneous presence of fairly large needles of the α′-phase. Inflows of deposited metal with a twice as large diffusion zone of the order of 20-40 μm are observed. The first row of grains of the deposited metal refers to the equiaxed ones, which is related to the conditions of heat removal. All subsequent rows of grains have a characteristic columnar structure. If the columnar grains in the structure of the transition region from the base metal had dimensions of about 80-120 X 200-300 (μm), then the grains in the sample closer to the center are about 2-3 times larger, and the structure itself is less homogeneous. It also contains inclusions of equiaxed configuration, which are horizontally formed columnar crystals. The growth of such grains occurs due to the movement of the crystallization front co-axially with the formation of the deposit. Such structures are observed closer to the existing solid phase of previously deposited layers.

For the studied alloy in the annealed state, the microstructure (Figure 1b) is characterized by a fine-crystalline, but inhomogeneous structure with the size of the α-phase colonies from 15 to 50 μm. The thickness of the α-phase plates varies from 1 to 5 μm. Despite some structural inhomogeneity, the distribution of alloy components is shown to be as uniform as possible, with the exception of large grain boundaries, in which a slightly lower concentration of niobium is observed.

After quenching, the macrostructure of the samples does not change, as indicated by the configuration of the macrograin boundaries. The microstructure (Figure 1c) is characterized by incomplete transformation. Inside the macrograins, a martensite-like acicular structure is formed, and closer to the boundaries there may be remnants of the lamellar structure of the untransformed α-phase with a length of 5 to 30 μm. This indicates an unevenly increased concentration of oxygen at the macrograin boundaries. As in the previous case, the distribution of elements shows a reduced concentration of niobium and an increased concentration of titanium at grain boundaries.

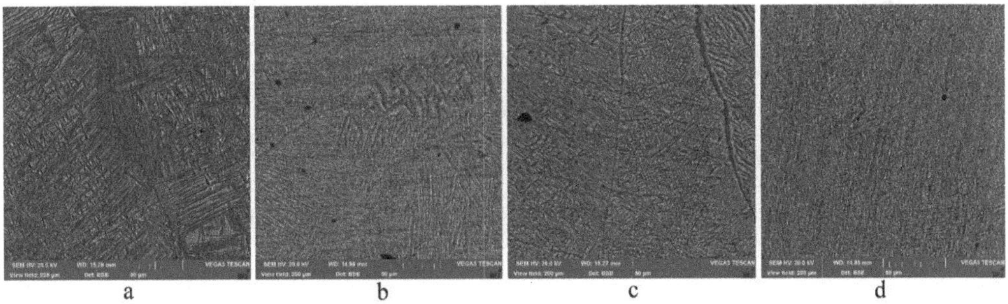

Figure 1. Microstructure of the samples (Magnification X1400): a – after surfacing, b – after vacuum annealing, c – after quenching, d – after quenching with aging.

The sample after quenching with aging has the most homogeneous structure (Figure 1d). In them, the partial disintegration of acicular martensite-like is noticable and the preservation of a thin lamellar structure near the oxygen-saturated areas of the grain boundaries.

Due to the high content of oxygen in the deposited metal, it is impossible to produce cylindrical samples for strength testing, since the deposited metal is very brittle and not amenable to lathe processing. It was also impossible to test flat samples made by the electrospark method. The fragility of the metal is also confirmed by the high values of microhardness - 4200-5000 MPa. The microhardness of the samples after various types of heat treatment is not particularly different and is at the level of 4430-4490 MPa.

4 CONCLUSIONS

Layer-by-layer TIG surfacing (WAAM method) with filler metal cored wire based on Ti-13Zr-13Nb alloy was performed. As a result, a 9-layer sample with a height of 15.2 mm and a width of 11 mm was obtained. There are almost no pores in the obtained sample by WAAM, the structure is homogeneous. The local chemical composition analysis showed the homogeneity of the distribution of elements: Zr = 6.5% by weight, Nb = 6.7% by weight. The number of these elements decreased due to the presence of a pure titanium shell in the metal powder wire. For all samples, the average microhardness values are very similar and differ by no more than 5%. At the same time, the microstructure of the samples in the initial state and after heat treatment differs significantly. After quenching with aging, the structure became more homogeneous, compared to other types of heat treatment. Due to this type of heat treatment partial disintegration of acicular martensite-like parts is noticable and the preservation of a thin lamellar structure near the oxygen-saturated areas of the grain boundaries disapears.

REFERENCES

Wang, K. 1996. The use of titanium for medical applications in USA. *Material Science and Engineering*. 213(1-2): 134–137.

Silva, H.M. et al. 2004. Study of nontoxic aluminum and vanadium-free titanium alloys for biomedical applications. *Materials Science and Engineering: C*. 24(5): 679–682.

Niinomi, M. 2008. Mechanical biocompatibilities of titanium alloys for biomedical applications. *Journal of the Mechanical Behavior of Biomedical Materials*. 1(1): 30–42.

Yoshimitsu, O. et al. 1993. Effect of alloying elements on mechanical properties of titanium alloys for medical implants. *Materials transactions JIM*. 34(12): 1217–1222.

Shneider, S. et al. 2005. Study of the Non-Linear Stress-Strain Behavior in Ti-Nb-Zr Alloys. *Materials Research*. 8(4): 435–438.

Sokolovskyi, M., & Bernatskyi, A. 2023. Developmental review of metal additive manufacturing processes. *History of science and technology*. 13(2): 334–356.

Ladokhin, S.V. et al. 2020. Features of Obtaining Titanium Alloys of Ti–Al–Si–Zr–Mo–Nb–Sn System under Conditions of Electron-Beam Foundry Technology. *Casting processes*. 140(2): 8–14.

Yanko, T.B. et al. 2018. Technology for obtaining of plasma spheroidised HDH titanium alloy powders used in 3D printing. *Technological systems*. 85(4): 36–41.

Schwab, S.L. et al. 2023. Welding materials for TIG welding, surfacing, and WAAM technology of titanium alloys. *Welding in the World*. 67(4): 981–986.

Akhonin, S.V. & Schwab, S.L. 2019. Filler flux-cored wire for tig welding and surfacing of VT22 titanium alloy. *The Paton Welding Journal*. 2019, (6): 34–37.

Ryabtsev, I.O. et al. 2023. Methods of Investigation of the Deposited Metal Properties and Their Application for the Development of Flux Cored Wires. *Materials Science*. 59(4), 467–473.

Session 2. Advanced technologies

of special electrometallurgy

Welding and Related Technologies – Krivtsun et al. (Eds)

New method for improving physical homogeneity of steel ingots

V.O. Shapovalov, F.K. Biktagirov, V.V. Barabash, O.V. Hnatushenko, O.M. Gnizdylo & I.V. Protokovilov
E.O. Paton Electric Welding Institute of the NAS of Ukraine, Kyiv, Ukraine

ABSTRACT: The problem of physical homogeneity of ingots has been a subject of study for several centuries and remains unresolved. The ingots have axial porosity and cavities that form during the metal solidification process. A new method for controlling the ingot solidification by applying a combined external effect is proposed that of heating the top of the ingot, combined with stirring the liquid metal in the inner middle part of the ingot during its solidification. The results from physical and numerical simulation confirm the ability of increasing the physical homogeneity of ingots. Research of quality of ingots, which had a height-to-diameter ratio of 2.1...2.5 and a taper of 3-5%, showed that the ingots were free of gross defects of shrinkage origin, and they had reduced axial porosity. The proposed method may be a promising one for increasing the physical homogeneity of large steel ingots.

1 INTRODUCTION

One of the main defects in steel ingots obtained by pouring liquid metal into molds is their physical heterogeneity in the form of voids, cavities, looseness, and reduced metal density.

The cause of such defects is the shrinkage of the metal, i.e., the reduction in its volume during solidification. As the mass of the ingots increases, the development of physical heterogeneity intensifies. This is largely due to the formation of an extensive two-phase area during the cooling process. In this zone, the flow of liquid metal to the solidifying areas is hindered, making it difficult to compensate for shrinkage. The highest porosity and looseness of the metal are observed in the upper axial zone of ingots, which solidifies last.

Currently used methods to improve the physical homogeneity of steel ingots, such as the improvement of pouring techniques, modification of the molds' geometric parameters, using of modern insulating materials in the ingot's top part, have largely been exhausted. These methods do not allow for a fundamental influence on the solidification conditions of the liquid core of the ingot.

Therefore, further improvement of the quality of steel ingots is possible due to the use of technological methods that allow influencing the conditions of formation of their internal structure. One such method could be a new combined method of external influence on the conditions of ingot formation, which involves heating the metal in the head part and stirring the liquid core of the ingot. The present work is dedicated to evaluating the possibility of increasing the physical homogeneity of steel ingots using the aforementioned method of combined influence.

2 METHODOLOGY AND EXPERIMENTS

Initially, physical and numerical simulation was carried out for a 50-ton sheet ingot with a height of 3000 mm and dimensions of the wide and narrow sides being 2000 mm and

DOI: 10.1201/9781003518518-4

1000 mm, respectively. This ingot was selected as the object of study because it exhibited shrinkage-related defects when cast in a mold.

Physical simulation was performed on a flat model scaled at 1:10, maintaining geometric similarity to the longitudinal section of the ingot. Sodium thiosulfate was used as the modeling medium, a substance widely employed in physical simulation of steel ingot solidification (Protokovilov et al. 2019).

Three scenarios of ingot production were simulated: without additional heating of the ingot head; with heating of the upper surface of the modeling medium; and with heating combined with periodic stirring of the unsolidified part of the model ingot. The temperature of the modeling melt was measured, and the solidification process was documented using photo and video recordings.

Numerical simulation was conducted using the Comsol Multiphysics software package version 6.1, focusing on the solidification of the same steel ingot made of grade Q345 steel. The positions of the liquidus and solidus isotherms were determined based on the temperature distribution within the liquid phase and the transition zone from liquid to solid state, taking into account the temperature dependencies of the metal's thermophysical properties. All forms of heat exchange with the surrounding environment were considered.

The modeling included scenarios of ingot production with heating of the metal in the top part and with the same heating combined with periodic stirring. The heating simulation assumed the presence of a medium with a temperature of 1650°C on the upper surface of the metal. Six hours after the start of ingot solidification, this temperature was reduced to the metal's liquidus temperature over a period of two hours. This simulated the temperature conditions that may occur during electroslag heating of the metal in processes like ESH and ESHT-J (Biktagirov 2002, Nakano et al. 2014).

Both physical and numerical simulation do not fully capture the complex processes that occur during metal solidification in real conditions. They provide a qualitative assessment, but nevertheless allow us to evaluate the difference in solidification conditions with and without the application of external influences.

After the aforementioned physical and numerical simulation, experiments were conducted to produce actual ingots. Under laboratory conditions, steel ingots weighing 120 kg were cast in molds made of refractory material. The use of a refractory mold made it possible to simulate the solidification conditions of larger ingots cast in cast-iron molds. This approach was also used by the authors of (Sang et al. 2010) when verifying the results of mathematical modeling on real ingots of relatively small mass.

Ingot casting was performed using steels of grades C22, C80W2, and 41Cr4. The ingots had diameters of 220-230 mm, with a height-to-diameter ratio of 2.1 to 2.5, and a taper of 3-5%. Several series of ingots were cast, both without external influence (control ingot) and with the application of heating and metal stirring (experimental ingots). Heating was performed using the electroslag method, and stirring was achieved by introducing a stirring device into the solidifying ingot. Axial templates were cut from the resulting ingots, and the macrostructure, density, and hardness of the metal were studied.

3 RESULTS AND DISCUSSION

The results of physical simulation showed that the application of additional heating and stirring leads to an increase in the angle α, which characterizes the geometry of the metal solidification front (Figure 1). Such a change in the geometry of the solidification front reduces the likelihood of closed cavities forming along the ingot's axis, where the melt lacks sufficient feed, thereby creating favorable conditions for the elimination of axial porosity.

According to numerical simulation data, applying metal heating in the 50-ton sheet ingot prevents the formation of a solid metal "bridge" at the transition level from the mold to the hot top. As demonstrated in the work (Biktagirov et al. 2019) through numerical simulation of the solidification of such an ingot using the "MagmaSoft" software package, the difficulty in the flow of liquid metal from the hot top to the lower levels is the cause of shrinkage cavities forming in the upper part of the ingot when using traditional method of its production.

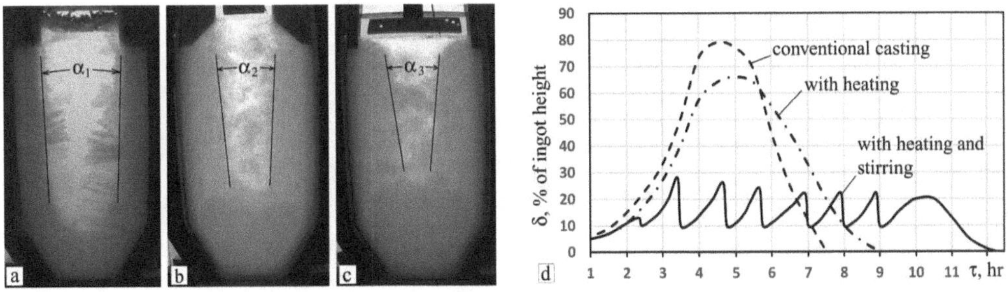

Figure 1. The effect of external influence on the conditions of ingot solidification: a – without heating, b – with heating, c – with heating and stirring ($\alpha_1<\alpha_2<\alpha_3$); d – change in the extent of the two-phase zone (d) along the axis of a 50-ton sheet ingot.

However, heating has little effect on the solidification conditions of the main part of the ingot. For a sufficiently long time, the dynamics of changes in the width of the two-phase zone (the distance between the liquidus and solidus lines) in scenarios without heating and with heating differ little from each other (Figure 1d). The restraining effect of heating on the growth of the two-phase zone becomes apparent only when the boundary of the solidification onset (the liquidus line) approaches the heated surface (the top of the ingot). Therefore, the size of the two-phase zone along the ingot's axis during heating is not significantly smaller than without heating (Figure 1d). At the same time, heating combined with stirring allows for a significant reduction in the size of the liquid-solid region during the metal's solidification process (Figure 1d). During stirring, the overheated metal from the upper part of the ingot is drawn inward, which leads to a rapid reduction in the width of the two-phase zone. This not only helps to prevent the formation of gross shrinkage-related defects in the ingot but also significantly reduces axial looseness and porosity in the metal.

The results of physical and numerical simulation show that the combined method of influencing the conditions of steel ingot formation by heating and stirring the metal creates conditions for increasing the physical homogeneity of steel ingots.

Studies of actual ingots confirm the above findings. Figure 3 shows photos of the macrostructure of axial templates cut from 41Cr4 steel ingots obtained without any influence on the solidification conditions (Figure 2a) and with the application of metal heating and stirring (Figure 2b). It is clearly visible that in the first case, the upper part has a pronounced area of increased porosity and looseness. In the experimental ingot, obtained with external influence, the metal in these areas is dense and free from visible defects.

In the control ingot, as seen from the data presented in Figure 3a, there is a sharp decrease in metal hardness in its upper part. This also indicates the presence of shrinkage-related defects. In the experimental ingot, the decrease in hardness in its upper part is minimal.

The significant improvement in the physical homogeneity of the steel ingot when using the considered method of external influence is evidenced by the metal density measurements (Figure 3b). In the ingot made of C22 steel, the metal density in its axial zone changes slightly from bottom to top – by only 0.115 g/cm³, which represents just 1.5% in relative terms.

The conducted studies and experiments show that one of the ways to improve the quality of steel ingots, cast by pouring liquid metal into molds, is to actively influence the metal solidification conditions. This can be achieved through a comprehensive method of external influence, which involves heating and stirring the metal during the formation of the steel ingot.

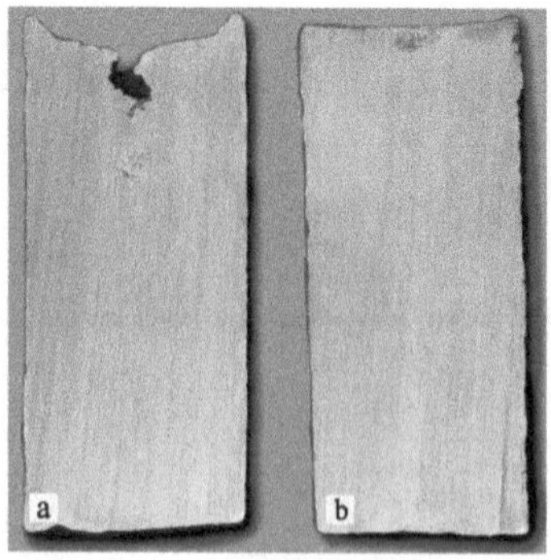

Figure 2. View of axial templates of control (a) and experimental (b) ingots.

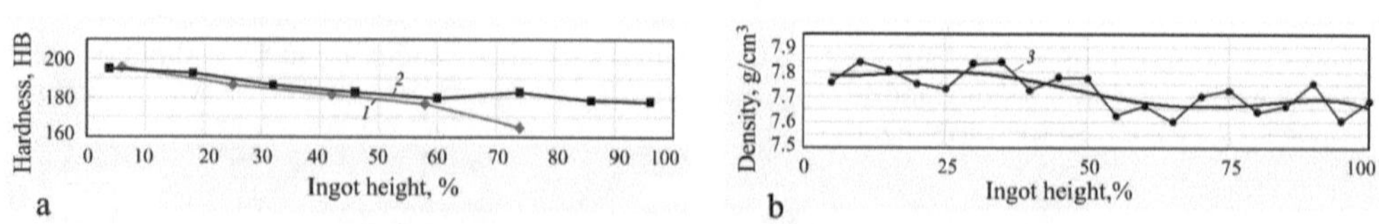

Figure 3. Change in hardness (a) and density (b) of metal by height of steel ingots: 1 – control ingot (steel 41Cr4); 2 – experimental ingot (steel 41Cr4); 3 – experimental ingot (steel C22).

4 CONCLUSIONS

1. A new combined method of external action was tested in obtaining steel ingots by pouring liquid metal into molds, which consists of heating and stirring the metal during its solidification.
2. According to physical and numerical simulation, this external action improves the conditions for ingot solidification by reducing the extent of the two-phase zone. This creates conditions to prevent the formation of significant shrinkage defects in the ingot and substantially reduces axial porosity and looseness.
3. It has been shown that in steel ingots obtained using metal heating and stirring, the density and hardness of the metal in their upper axial part are significantly higher compared to ingots produced without such external action.
4. The method of heating and stirring the metal may be promising for improving the quality of large steel ingots.

REFERENCES

Biktagirov, F.K. 2002. Application of electroslag process with non-consumable electrodes for melting, refining and treatment of metals. Part 1. *Electrometallurgy Today 4*, 9–14.

Biktagirov, F.K., et al. 2019. Computer modelling of 50-ton sheet ingot forming process. *Metal and casting of Ukraine 5-6*, 1–7.

Nakano, M. et al. 2014. Manufactory of trial rotor forging of 9% Cr steel containing Co and B (X13CrMoCoVNbNB9-2-1) for ultrasupercritical stream tuhbines. *Advances in Materials Technology for Fossil Power Plants: Proceedings from the Seventh International Conference*, October 22-25, 2013, *Waikoloa, Hawaii, USA*. 321–332.

Protokovilov, I.V. et al. 2019. Physical modeling of ingot crystallization in the mold under the conditions of electroslag heating and topping. *Electrometallurgy Today 3*, 3–9.

Sang, B.G. et al. 2010. Study on macrosegregation in heavy steel ingots. *International Journal of Cast Metals Research. .23(4)*, 205–210.

Welding and Related Technologies – Krivtsun et al. (Eds)

Electroslag remelting of variable cross-section electrodes using a two-circuit scheme

Ye.O. Pedchenko, V.L. Petrenko, Yu.V. Kostetskyi, G.O. Polishko & V.A. Zaitsev
E.O. Paton Electric Welding Institute (PWI), Kiev, Ukraine

ABSTRACT: The possibility of using electroslag remelting to recycle spent rolls from cold rolling mills and continuous casting machines is investigated. The expediency of remelting rolls without cutting off the necks and welding into a solid consumable electrode is shown. Experimental melts were used to investigate the use of ESR in a two-circuit scheme for remelting ingots of variable cross-section, which imitate rollers with necks. During the experimental melts, the geometry of the electrodes provided a change in the filling factor in a wide range from 0.69 to 0.2. It was experimentally confirmed that it is possible to surfacing an ingot while maintaining a constant cross-sectional profile of the liquid metal bath despite a significant change in the value of the filling factor during remelting. The mathematical modeling of the current distribution in the slag bath at ESR of electrodes of variable cross-section was carried out and the operating parameters of the remelting process were determined to obtain an ingot with a constant profile of the liquid metal bath along its cross-section. It has been established that the optimal conditions for preserving the remelting speed and depth of the two-phase zone are achieved when the electric power supplied to the consumable electrode exceeds the power supplied to the current-carrying mould by a factor of two.

Keywords: electroslag remelting, recycling, rolls, continuous casting machines, consumable electrode, two-phase zone, current-supplying mould, two-circuit scheme

1 INTRODUCTION

The Electric Slag Remelting (ESR) process is widely used in the world to recycle spent rolls from cold rolling mills and rolls from continuous casting machines. For example, at NKMZ in Kramatorsk, Ukraine, there has been a practice of remelting spent rolls by ESR followed by the production of new forged rolls from ESR metal. The ESR process helps to achieve a fine and homogeneous dendrite structure, which is essential for the durability and performance of these rolls (Sato 2011). The expertise of Sheffield Forgemasters, England, in combining ESR with casting techniques for the recycling and manufacture of cold rolled rolls is another example. Overall, there is now no doubt that the use of ESR in the recycling of used rolls from cold rolling mills and rolls from continuous casting machines is technically and economically sound.

However, there are some difficulties associated with direct electroslag remelting of scrap rolls. In classical ESR technology, it is necessary to ensure a constant mould filling factor to obtain a quality ingot, since this influences the quasi-stationary melting mode of the electrode, which directly affects the quality of the obtained ingot. However, a typical roll has a variable cross section whose diameter changes almost twice. This will significantly affect its melting conditions when remelting a roll as it is.

During the ESR of an electrode of variable section, which is a roll, its melting mode will change significantly, and accordingly, the conditions of ingot crystallization. During the remelting of a variable section electrode at a constant power input, the mass melting rates are greater at the maximum filling factor on the barrel and the lowest on its neck. Accordingly,

DOI: 10.1201/9781003518518-5

the ingot deposition rate will change. The latter in turn affects the depth and shape of the liquid metal bath, the parameters of which determine the direction of growth of dendrites and their size. In order to obtain an ingot of uniform height quality, it is necessary to ensure a constant ingot deposition rate and a corresponding stable depth of liquid metal bath.

Therefore, to apply ESR according to the classical scheme, it is necessary to form a consumable electrode of constant diameter from waste rolls to ensure a constant mould filling factor. For this purpose, it is necessary to cut the necks of the rolls and weld their barrels together, and then to weld the inventory head to the obtained electrode. Typically, an electrode consisting of 2-3 barrels welded together is required to make a new roll from an ESR ingot. All these preparatory processes take time and require additional investment in electrode preparation, which reduces the economic efficiency of recycling and increases its time.

The modern dual-circuit ESR technology allows to break the close link between the mould filling factor and the electrode melting mode, and to obtain ingots of higher quality compared to the classical remelting scheme (Tsykulenko 2000). Special current-supplying moulds (CSM) are used to perform the ESR process in a two-circuit scheme. In the case of ESR CSM, the power is supplied by two circuits - "consumable electrode – bottom plate" and "mould – bottom plate" (Medovar 2005). In this scheme, the current-supplying mould acts as a non-consumable electrode. It provides opportunities to break the rigid link between temperature parameters and the performance of the ESR process, and significantly expands the possibilities of controlling the thermal processes occurring in the slag and metal bath, which was hardly achievable in the classical ESR (Kuskov 1996, Kuskov 2022). The temperature conditions of the slag bath in the ESR CSM can be controlled within certain limits regardless of the power on the electrode. This increases the ability to control the volume and shape of the metal bath and the length of the two-phase zone, thereby influencing the quality of the metal (Kuskov et al. 1995). Thus, this technique can be used for remelting rolls by using them as consumable electrodes without their preliminary preparation. As follows the use of ESR CSM technology will greatly facilitate and reduce the cost and speed up the process of roll recycling.

The purpose of the study was to investigate the peculiarities and to determine the rational method of remelting used rolls by ESR CSM technology without their special preparation by cutting the necks of the rolls and obtaining an ingot of uniform quality in terms of its height.

2 RESEARCH METHODOLOGY

In order to model the recycling of rolls, the ESR CSM of an electrode with a variable cross section was performed. The electrode was composed of three distinct sections, each with a varying diameter. This was done to create a simulation of the electrodes from real rollers. The diameters of 150, 110, and 80 mm were utilized to represent the roller barrel, the transition from the barrel to the neck, and the neck of a genuine roller from a continuous casting machine, respectively. During the remelting process, the electrode's distinctive geometry ensured a significant change in the filling factor, from 0.69 to 0.37 to 0.2, for the mould with a forming part diameter of 180 mm. Each section of the model electrode weighed 28 kg.

Remelting was carried out according to the two-circuit scheme on the modernized ESR unit type of P-951 with two power sources (72V/10kA each) and thyristor voltage regulators. The electrode and the current supply part of the mould were connected to separate transformers, which made it possible to control the voltage supplied to each circuit separately. A short, widened current supply mould with the diameter of the widened part being 225 mm and the diameter of the forming part being 180 mm was used for experimental remelting.

In order to ascertain the configuration of the liquid metal bath, iron sulfide was introduced to the melting zone at specific intervals during the remelting process. Longitudinal templates were cut from the obtained ingots and then made their sulfur prints to observation the shape of the liquid bath.

Mathematical modeling of the electric current distribution in the slag bath was performed in the geometric form of the slag bath as close as possible to the real experiment. The specific electrical conductivity of the slag was assumed to be 150 S/m, and the specific electrical conductivity of the electrode, ingot and current-carrying section was assumed to be 56.10^6 S/m.

3 RESULTS AND DISCUSSION

In the first experiment, it was necessary to determine the limits of the influence of changes in the power supply to the current supply section of the mould on the shape of the liquid metal bath.

It was assumed that the increase in the power of the current supply section would compensate for the power supplied to the electrode, and the estimated remelting performance should be constant. However, as it turned out, the effect of the current supply section at a minimum filling factor of 0.2 no longer ensured the maintenance of productivity. Note that the filling factor of 0.2 is critically low for the classical ESR scheme (Medovar 2020). The remelting parameters obtained in the first experiment are presented in (Table 1). In this experiment, an ingot with a diameter of 178 mm and a weight of 84 kg was obtained, the length of the ingot was 432 mm (Pedchenko 2022).

Table 1. Remelting parameters of an electrode with variable cross section in Experiment No. 1.

	Unit of measurement	Filling factor 0,69	0,37	0,2
P_{el}	kVA	100	100	120
P_{csm}	kVA	0	50	120
Melting rate	kg/h	140	120	110
Depth of liquid metal bath	mm	80	50	30

The study of the templates of the experimental ingot cross section showed that the depth of the liquid metal bath decreased from 80 mm to 30 mm as the power supplied to the current supply section of the mould increased. At the same time, the remelting productivity decreased from 140 kg/h to 110 kg/h.

For a more detailed study of the influence of the power distribution between circuits, mathematical modeling was performed to estimate the current density at the periphery of the ingot based on the experimental conditions and to adjust the melting parameters so that the depth of the liquid metal bath is constant during the remelting process. The electrical parameters of the natural experiment were transferred to the mathematical model. The simulation results clearly show that the current density at the periphery of the ingot differs by a factor of 2. Thus, with a filling factor of 0.69, the current density at the periphery was 0.8-1 (104 A/m^2), with a filling factor of 0.37 1.2-1.4 (104 A/m^2), with a filling factor of 0.2 1.4 -1.6 (104 A/m^2). Accordingly, the amount of heat generated in the slag bath in this zone will increase - which affects the profile of the molten metal bath. Considering the influence of the current supply section of the mould, it was determined that the performance of the remelting process should be ensured by the power supplied to the electrode, since the effect of the current supply section on productivity is not significant, but it strongly affects the depth of the liquid metal bath.

Based on the results of the mathematical modeling, the second full-scale experiment was conducted, in which the melting of the ESR CSM electrode of variable cross-section was performed according to the calculated data. The melting parameters were adjusted so that the current density on the periphery of the ingot was the same during the entire remelting - about 1.2 (104 A/m^2).

Table 2 shows the results of the experiment carried out on the adjusted parameters, which data agree with the calculated data regarding the effect of the current supply section of the mould on the depth of the liquid metal bath. The possibility of ensuring constant productivity by means of the power supplied to the electrode also proved to be true.

Table 2. The remelting parameters of the variable-section electrode of Experiment No.2.

	Unit of measurement	Filling factor 0,69	0,37	0,2
P_{el}	kVA	80	120	200
P_{csm}	kVA	50	50	70
Melting rate	kg/h	120	120	120
Depth of liquid metal bath	mm	50	45	48-45

From the result of the full-scale experiment (Table 2), we can see that with the increase of the power at the current supply section of the mould the depth of the liquid metal bath decreased only from 50 mm to 45 mm, which is within the measurement error of the bath shape fixation method by iron sulfide. The ESR CSM process enables to exert influence not only on the profile of a liquid metal bath, but also to ensure a reduction in its depth by a factor of almost two.

4 CONCLUSIONS

The potential of electroslag remelting as a means of recycling spent rolls from cold rolling mills and continuous casting machines is investigated. The feasibility of remelting rolls without severing the necks and welding them into a single, indivisible consumable electrode using ESR CSM technology is demonstrated. This approach will facilitate and reduce the cost and speed up the process of spent rolls recycling.

The experimental results demonstrated that it is feasible to remelt an ingot while maintaining a constant cross-sectional profile of the liquid metal bath, despite a significant change in the value of the filling factor, by employing ESR CSM technology with a two-circuit scheme. During the experimental melts, the geometry of the electrodes permitted a change in the filling factor to vary significantly, from 0.69 to 0.2.

It was determined that the performance of the remelting process should be ensured by the power supplied to the electrode, as the effect of the current supply section of the mould on productivity is not significant, but it strongly affects the shape of the liquid metal bath. The ESR CSM process enables the profile of a liquid metal bath to be influenced, thereby reducing its depth and ensuring its stability and shape during the remelting of an electrode with a variable cross-section. In order to maintain a stable depth of the liquid metal bath in the case of ESR CSM electrodes of variable section in a mould, it is necessary to perform a process with constant productivity and current density at the periphery of the liquid slag bath.

In consequence, the ESR CSM process enables the profile of a liquid metal bath to be influenced, thereby reducing its depth and ensuring its stability and shape during the remelting of an electrode with a variable cross-section. In order to maintain a stable depth of the liquid metal bath in the case of ESR CSM electrodes of variable section in a mould, it is necessary to perform a process with constant productivity and current density at the periphery of the liquid slag bath.

REFERENCES

Sato J. At all. 2011. High-quality Work Roll Manufacturing Technology Using New Electro Slag Remelting (ESR). Kobelco Technology Review. No. 30: 7–12.

Tsykulenko A.K. et al. (2000) Two-circuit circuit of electroslag remelting of consumable electrode. Probl. special Electrometallurgy, 3, 16–20.

Medovar, L. et. al. (2005) ESR with two power sources and process control. In: Proc. of Int. Symposium on Liquid Metal Processing and Casting (Santa-Fe, New Mexico, USA, Sept.11–14, 2005), 131–135.

Kuskov Y.M., Us S.V., Tomilenko S.V. 1995. Investigation of the parameters of electroslag melting in a current-supllying mould. Problems of special electrometallurgy. No. 3: 24–28.

Kuskov Y.M. at al. 1996. Current distribution in a slag bath during electroslag surfacing with granular filler material. Automatic Welding. No.11: 6–11.

Kuskov Y.M. at al. 2022. Current-conducting mould in electroslag technologies. Automatic Welding. № 7: 40–45. DOI 10.15407/as2017.12.05

Medovar L.B. at al. 2020. Electroslag remelting of consumable electrodes of variable section. Electrometallurgy Today. №4: 3–6. DOI 10.37434/sem2020.04.01

Pedchenko E.O., at al..2022. Recycling of used noncompact tools from high-speed steel by electroslag remelting. Electrometallurgy Today №3: 23–28. DOI 10.37434/sem2022.03.04

Welding and Related Technologies – Krivtsun et al. (Eds)

Evolution of liquid-phase methods for growing molybdenum single crystals

V.O. Shapovalov, V. Yakusha, Y. Nikitenko & I.V. Protokovilov
E.O. Paton Electric Welding Institute of the NAS of Ukraine, Kyiv, Ukraine

X.S. Liu & H.B. Pan
State Key Laboratory of Precision Welding & Joining of Materials and Structures, Harbin, China

ABSTRACT: The evolution of liquid-phase methods for growing molybdenum single crystals is analyzed. It is shown that the methods for growing crystals with a single heating source cannot solve the problem of producing industrial-sized crystals. Super-large molybdenum crystals of various configurations can be produced using the combined plasma induction method. This method is already used for growing single-crystal ingots, as well as single-crystal plates. The monocrystalline quality of the ingots and the preservation of the orientations of the selected crystallographic axes were confirmed by the X-ray method according to Laue. Single crystals of refractory metals with a given orientation have prospects for application in the power optics of high-power lasers.

1 INTRODUCTION

In the 50s and 70s of the twentieth century, the following unique properties of single crystals of refractory metals were revealed: high plasticity, absence of gas emission, resistance to recrystallization, creep, action of plasma and alkali metal melts, radiation and thermal cyclic effect, special mechanical properties, etc. The studies were carried out with single crystals of various purities, including both high-purity crystals and crystals of technical purity. The use of molybdenum single crystals was forecast in the following industries: X-ray technology (anti-cathodes); electrical industry (contacts and wires); laser technology (mirrors for optical and X-ray lasers); nuclear and thermonuclear research installations (devices for channelling particles, etc.).

However, the high expectations for the widespread use of molybdenum single crystals in industry, aerospace, and military technology have not yet been realized in practice. The main reason is the small size of single crystals, typically in the form of rods up to 25 mm in diameter, making it difficult to manufacture large parts or structures. Today, large molybdenum single crystals are needed to manufacture power optics elements for super-powerful lasers. In this regard, the prospect of growing large single crystals in the form plates as billets for large-format single-crystal rolled products is particularly interesting. The answer to it can be obtained in the process of analyzing liquid-phase methods for growing tungsten single crystals.

2 TRADITIONAL LIQUID-PHASE METHODS FOR GROWING MOLYBDENUM SINGLE CRYSTALS

The main liquid-phase method for growing tungsten, rhenium, tantalum, molybdenum single crystals is a crucible-free zone melting. As a heating source, an electron beam was originally used (Calverly et al. 1957, Glebovsky 2019). One of the first options for implementing this method is shown in Figure 1a. The molten zone is the main link in the process of converting a polycrystal

DOI: 10.1201/9781003518518-6

into a single crystal. The method is unique for growing small-diameter single crystals. The crystals obtained by this method were used both for research purposes and for manufacturing of various parts of measuring instruments and scientific equipment. Attempts to sig-nificantly increase the

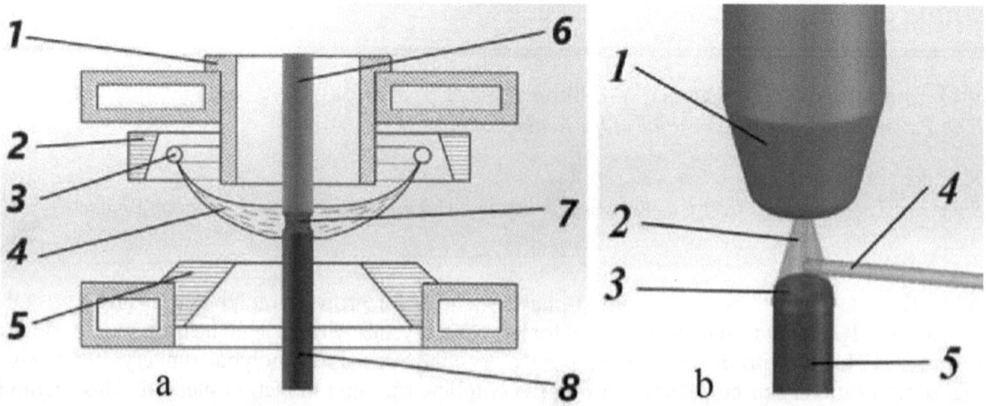

Figure 1. Methods for growing single crystals of refractory metals from the melt using a single heating source: a – electron beam zone melting (1 – upper accessory electrode; 2 – focusing electrode; 3 – cathode; 4 – circular electron beam; 5 – lower accessory electrode; 6 – polycrystalline specimen (anode); 7 – liquid zone; 8 – single crystal); b – plasma-arc growing method (1 – plasmatron; 2 – plasma-arc; 3 – liquid zone (metal pool); 4 – consumable rod; 5 – single crystal).

diameter of single crystals failed. For increasing the crystal diameter, it was necessary to increase the power of the electron beam gun. That led to an increased height of the molten zone and overheating of its surface layer, which resulted in a decreased viscosity of the melt and an increased hydrostatic pressure in the lower part of the melt. The zone lost its stability, and the metal spilled.

Different types of zone melting using electron beam heating have been developed: zone compression melting and peripheral zone melting. The practice of growing cylindrical single crystals of molybdenum has shown that it is impossible to grow crystals of large diameter of more than 30 mm using these methods (Bucklow & Wilson 1968).

The development of plasma-arc heating sources made it possible to create a new method for growing tungsten and molybdenum single crystals, which was similar to the Verneuil method (Figure 1b) (Savitsky & Burkhanov 1978). A characteristic feature of this method is that with an increase in the diameter of a single crystal, it is necessary to significantly increase the current of the plasmatron, which result in overheating of the pool surface. At the same time, the pool completely covers the end face of the single crystal and with a slight instability of the process it is very difficult to keep it from spills.

Analysis of these methods shows that the use of a single heat source does not allow solving the main problem – the problem of growing large single crystals. With an increase in the size of the crystal, problems arise associated with overheating of the metal pool, keeping the pool from spills, and high thermal stresses in the single crystal (Shapovalov et al. 2016).

3 PLASMA-INDUCTION METHOD FOR GROWING MOLYBDENUM SINGLE CRYSTALS

The plasma-arc technology of growing single crystals of molybdenum was further developed with creation of a combined plasma-induction method with a droplet transfer of liquid metal into a pool (Paton et al. 2016). The scheme of the method is shown in Figure 2a. Integration of two independent heating sources made it possible to obtain a new effect - grow not only large, but also profiled molybdenum single crystals, for example, in the form of plates. The method has several fundamentally different characteristics compared to the methods described above: 1) - flexible control of the thermal state of a single crystal is provided by a high-frequency heating source. As a result, the level of stresses in the crystal decreases, which number of dislocations decreases. The use of a high-frequency heating source made it possible not only to heat up the crystal, but also to keep the pool from spills due to levitation forces; 2) - the crystal is formed bymoving the local metal pool. When the pool moves under the influence of a plasma arc, a consumable polycrystalline rod is remelted. Thus, crystal formation occurs is done layer by layer. The metal in the form of drops enters the local metal pool. Usually, the diameter of the pool is in the range of 20...25 mm, and the depth of the pool ranges from 2.0 to 2.5 mm. With an increase in the diameter of the consumable polycrystalline rod, the mass of the droplets as well as the range of changes in the depth and diameter of the pool increases.

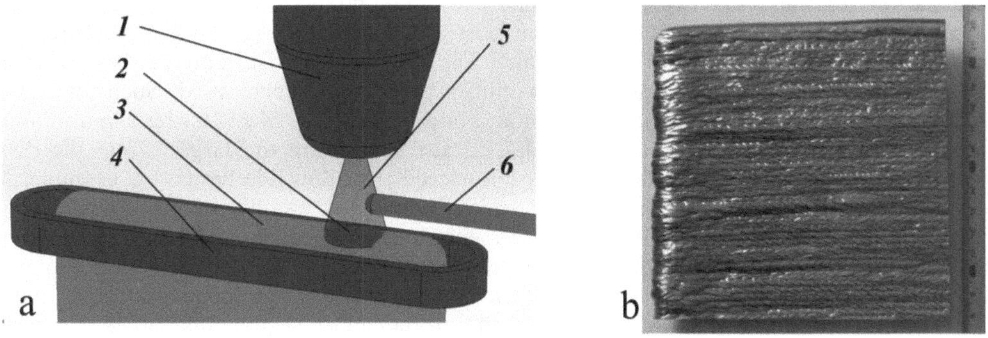

Figure 2. Plasma-induction method for growing molybdenum single crystals: a – diagram of the growing process (1 – plasmatron; 2 – liquid zone (metal pool); 3 – single crystal; 4 – inductor; 5 – plasma-arc; 6 – consumable rod); b – molybdenum single crystal with a size of 160×170×20 mm.

The use of two sources of heating made it possible to actively influence the shape of the interface between the solid and liquid phases. The pool with the plasma-induction method has a flatter shape than with the plasma-arc method. If phantom grains are formed in the surface layer, they do not grow inward, but are pushed out onto the crystal surface.

A small local pool, whose trajectory of movement can be controlled (3D technology), in combination with two heating sources, allows growing tungsten single crystals of theoretically unlimited sizes and shapes: flat ingots for large-format single-crystal rolled products, pipes, crucibles, rods of large diameters, etc.

4 RESULTS AND DISCUSSION

The results of growing single crystals by traditional methods are widely presented in the literature and are well known: small crystal diameter, fairly perfect structure, high purity, etc.

Of greatest interest is the 3D plasma induction technology for growing large single crystals of molybdenum, which has great prospects for increasing the size and variety of shapes of single crystals. A real technology has already been developed and implemented for growing plates with a size of 160×170×20 mm (Figure 2b). The technology solves the main problem – increasing the size of single crystals.

Interesting are the results of studying single-crystal wafers grown from commercially pure molybdenum. Metal refining in the process of growing single crystals is insignificant (Table 1).

The ingots have a slight deviation from the shape of an ideal plate. Thus, the deviation from parallelism of the largest edges was not bigger than 2 mm.

Table 1. Chemical composition of single-crystal molybdenum plates.

	Non-metal impurities, $\times 10^{-5}$ % (wt.)				Metal impurities, $\times 10^{-5}$ % (wt.)						
Material	O	N	H	C	Al	Fe	Ni	Si	Ca	W	Mo
Source material	20	-	-	40	20	30	20	30	30	300	base
Single-crystal Mo	10	20	2	30	18	25	20	30	25	300	base

The most important characteristic of the obtained molybdenum ingots is their structural perfection. The monocrystalline quality of the ingot and the preservation of the orientations of the selected crystallographic axes were monitored by the X-ray method according to Laue. X-ray methods make it possible to evaluate the single crystallinity of a small sample with an area of up to 5 cm^2. The area of the large plate being evaluated is at least 100 times larger than the capabilities of the equipment used. Such an ingot can be considered as a solid monocrystalline body in the case when during a multipoint X-ray examination, for example, of the lateral surface having the largest area, the disorientation of the subgrain structure does not exceed the allowable limits, for example, 5 angular degrees or 0.087 radians. To determine the monocrystallity of the ingot, on its lateral surface along its length and height, epigrams were taken at points located at intervals of 10-15 mm (20-30 points). The identity of the epigrams (indices of the plane and directions in it) served as a confirmation that the ingot was a monocrystalline body. An additional check consisted in repeating the aforementioned procedure while rotating the sample through an angle of 3.14 rad. These epigrams were also "identical". At the same time, the spread of deviations from the direction of growth did not practically exceed 0.055 rad (Paton et al. 2016).

The fully grown crystal adopts the orientation of the seed crystal. By comparing the X-ray images and Laue reflexes of the "as-grown" crystal (without subsequent heat treatment) and the seed crystal, it was determined that they are practically identical. The X-ray diffraction patterns for each face showed nearly identical features, confirming that the ingot is a single crystal body without significant crystal structure defects.

The results of the study of single crystals suggest that the plasma-induction method is a promising one for growing large molybdenum crystals. Real equipment makes it possible to significantly increase the size of single crystals. Research is currently underway to intensify the process of growing flat single crystals in order to reduce their cost.

5 CONCLUSIONS

Analysis of the development of methods for growing single crystals for more than 70 years has shown that it is impossible to grow large single crystals of molybdenum using a single source of high-temperature heating and a large metal pool. The problem is solved by the use of at least two heating sources and the use of a local movable metal pool – transformation of 3D technology. It can be assumed that when replacing the plasma-arc heating source in the plasma-induction method of growing with an electron-beam source, it will be possible to create a shallow flat pool equal in area to the cross-section of a single crystal. At the same time, the perfection of crystals, the efficiency of the process can be increased and the cost can be reduced.

With the development and refinement of the plasma-induction method of growth, new opportunities are emerging today for expanding the production and application of single crystals of refractory metals.

ACKNOWLEDGMENTS

This work was carried out with the assistance of the National Research Foundation of Ukraine, project № 2023.04/0030. We also express our gratitude for the preparation of materials to international partners: X.S. Liu, H.B. Pan (Harbin, China).

REFERENCES

Bucklow, I. & Wilson, J. 1968. A fascia technique for float-zone metal crystals. *J. Phys. E: Sci. Instrum.* (1): 363.

Calverly, A. et al. 1957. The floating – zone melting of refractory metals by electron bombardment. *J. Sci. Instr.* 34(4): 142–147.

Glebovsky, V. 2019. *Crystal Growth.* London: IntechOpen Lim.

Paton, B. et al. 2016. *Plasma-induction growth of profiled ingle crystals of refractory metals.* Kyiv: Nauk. dumka. *(in Russian).*

Savitsky, E. & Burkhanov, G. 1978. Grown of single crystals of high melting metal alloys and compounds by plasma heating. *Journal of Crystal Growth.* 43(4): 457–462.

Shapovalov, V. et al. 2016. Application of additive technologies for growing large profiled single crystals of tungsten and molybdenum. *The Paton Welding Journal.* (5-6): 134–136.

Welding and Related Technologies – Krivtsun et al. (Eds)

Features of the formation of a directional structure of complex-profile castings

K.H. Kvasnytska, Y.H. Kvasnytska, Z.V. Parkhomchuk, V.O. Noha, V.I. Veis & P.B. Kaliuzhnyi
Physico-technological Institute of Metals and Alloys, Kyiv, Ukraine

ABSTRACT: The thermophysical features of the formation of a directional structure of complex-profile castings (a turbine blade for a stationary gas turbine engine) from the SM88Y alloy in the UVNK-8P vacuum plate with variations in the crystallization rates Vcr. = 5; 10 mm/min were studied experimentally and analytically. The dependences of the change in the temperature gradient at the crystallization front and the speed of movement of the front on its position in the crystallizing casting were obtained. Metallographic analysis shows that with selected optimal crystallization parameters, a regular dendritic structure is formed along the entire length of the product, including transition section.

1 INTRODUCTION

Increasing the resource and reliability of gas turbine engines (GTDs) is associated with the need to increase operating temperatures at the turbine inlet, additional dynamic and vibration loads on structural parts. This causes an increase in the requirements, first of all, for the working blades of the turbines, which experience force loading and erosion-corrosion destruction in contact with an aggressive gas flow for a long time (Dexin 2018). The problems of long-term high-temperature strength of nickel alloys are based on the fact that the initiation of fatigue cracks in these materials occurs in the intergranular space, therefore it is desirable to have a blade with a minimum number of grains and the ability to control their crystallographic orientation (Balitskii et al. 2024).

To achieve a regular oriented structure in the process of directional crystallization, it is important to create such conditions in which the rate of crystal growth will be maximum, and the rate of nucleation - minimum. This allows to achieve the desired alignment of the crystal structure and improve the quality of the final product (Glotka et al. 2021).

High-precision cast blades are produced by pouring into a ceramic mold. For the production of ceramic forms, a suspension of ethyl silicate is traditionally used as a binder, but its production requires the use of organic solvents such as acetone or ethyl alcohol which are explosive and flammable. (Ivaskevich et al. 2021). In the production of ceramic shell molds based on ethyl silicate, the evaporation of alcohol from the tanks containing the slurry and drying shells requires the installation of equipment to collect and reduce emissions of harmful substances. Water-based shell mold production methods are devoid of these environmental and transport disadvantages. Preliminary studies indicate that the tested samples of corundum-based ceramics based entirely on the water system with the addition of Remasol showed a higher level of open porosity in the raw state, after heating to 950 °C and after baking. Specifically, the open porosity increased from 7.51% to 12.23%, which is 5.62% more than that of samples using a combined alcohol system or based on ethyl silicate (Wisniewski et al. 2020). This high level of open porosity helps eliminate problems with underfilling when pouring the melt into the mold due to better "breathing" of the ceramic shell molds. Thus, the use of

DOI: 10.1201/9781003518518-7

water-based binder technology improves the surface quality of castings. Fire safety and emissions cleanup issues are solved because water-based binders do not contain solvents or volatile substances.

In order to obtain high-quality castings from heat-resistant corrosion-resistant alloys, the task of these studies is to research the influence of temperature-speed conditions of directed growth on the features of the formation of the directed structure of castings of complex profile turbine blades of power gas turbines using ceramic molds from the ceramic shell is formed taking into account modern environmentally friendly technologies.

2 METHODOLOGY

The experiments were carried out with the aim of establishing the optimal modes of directional crystallization and improving the technological process of obtaining a regular oriented structure in the castings of the blades of the II stage of the UGT 5000 gas turbine engine (engine power 5000 MW, manufacturer of SE GRPC "Zorya" - "Mashproekt", Mykolaiv, Ukraine). For research, a series of heat and corrosion-resistant nickel-based alloys CM88 was used, the chemical composition of which corresponds to the standard (Instruction, 2011). Determination of the amount of the main components of alloys and impurities was carried out by a chemical method according to standard methods, microalloying additives were controlled by a chemical-spectral method with a relative error of ± 0.001%. The carbon content was determined by an analyzer of the company "Leco" (USA). The microstructure of the samples was investigated on a Tescan Mira 3 LMU scanning electron microscope, main characteristics: spatial resolution: 1 nm 30 kV, 2 nm 3 kV; - accelerating voltage: 200 V - 30 kV; - operating pressure in the chamber: high mode vacuum ≈ 9*10-3 Pa; - low vacuum mode 7 – 150 Pa.

The experiments used the developed technological process of obtaining one-off models of gas turbine blades using 3D printing. Production of disposable models from PLA polylactide was carried out on a Klema 3D printer with a step of 0.2 mm according to the given drawing. The ceramic mold was made according to the traditional technology of casting models, which are melted, by immersing the model block in a suspension followed by sprinkling with electro corundum. The thickness of the ceramic mold is 10-12 mm. The surface layer was applied with a suspension without crumbling to create a smooth surface of the mold. A special feature of the mold manufacturing process was the use of a water-based binder of the Remasol brand (silica). The preparation of the suspension on the water-based binder was carried out by mixing its initial components, without cooling, with the help of a mixer at a rotation frequency of the mixer impeller of 1000-2800 rpm. Firing is carried out at a temperature of 900-950 °C without a supporting filler for 4-6 hours.

Melting and pouring were carried out in a UVNK-8P vacuum furnace, which implemented the crystallization process according to the Bridgman-Stockbayer method. The metal pouring temperature is 1550-1570 °C. The local values of the parameters of the directional crystallization process during melting were calculated from the temperature distribution in the volume of the castings, which was measured experimentally at five points along the axis of the casting with the help of tungsten–rhenium thermocouples (BP 5/20, electrode diameter 0.30 mm).

3 RESULTS

In order to establish the influence of the technological parameters of the crystallization processes and, above all, of the ceramic molds made using the new technology, a series of experiments were conducted with the variation of the crystallization speed of the castings at two levels: 5 and 10 mm/min. The research was carried out on molten standard samples of the alloy (Figure 1a). Visual inspection of the outer and inner surface of heat-resistant alloy castings and metallographic studies revealed no micropores, clogging, shells, non-metallic inclusions after pouring into molds obtained by improved technology. (Figure 1b).

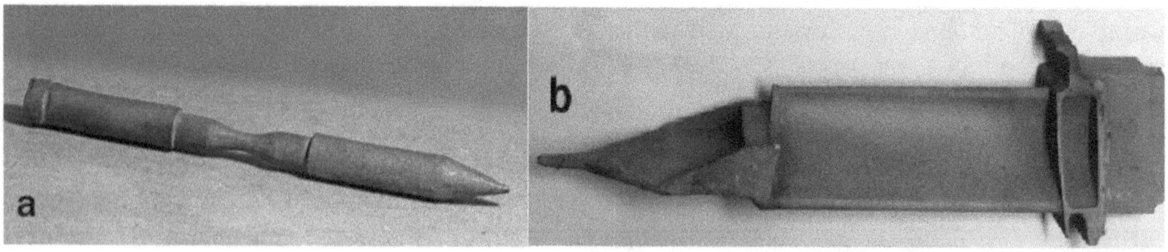

Figure 1. Appearance of the experimental sample (a); blades of the 2nd stage of the GTE turbine (b).

The macrostructure of test samples for mechanical tests is directionally crystallized (Figure 2). The microstructure of the alloy is typical for CM88-VI in the cast state: γ-solid solution, carbides, borides, and intermetallics. Quantitative characteristics of the dendritic structure - the distance λ between the dendritic axes of the 1st and 2nd order, associated with the cooling rate of the melt at the crystallization front, changed when the values of the velocities were varied - from λ= 220-165 μm at V=5 mm/min to 160 -170 μm at V =10mm/min. (Figure 3).

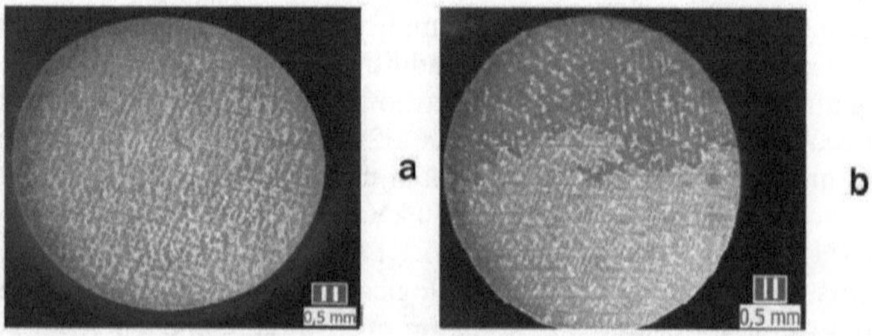

Figure 2. Macrostructure of samples (a. – one grain; b. – two grains).

It was established that with an increase in the crystallization rate of castings from 5 to 10 mm/min, the dendritic structure is refined, the volume content of the intermetallic strengthening γ′ phase increases with a simultaneous decrease in the size of its particles and the average cross section of the γ//γ′ - eutectic areas, as well as the amount of MC-type dispersed carbides in the interdendritic spaces increases.

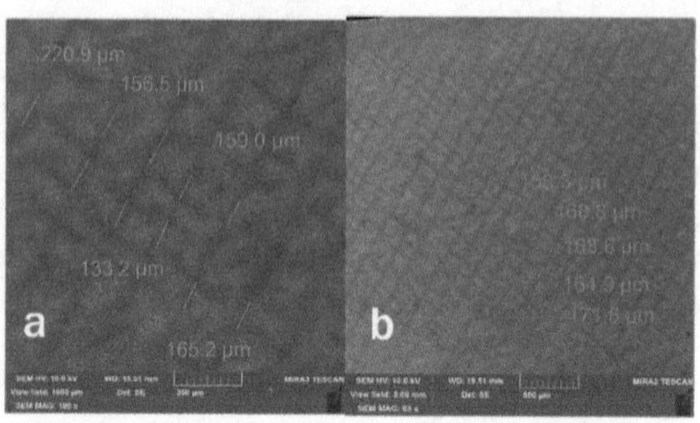

Figure 3. Microstructure of the cross-section of a sample of heat-resistant corrosion-resistant alloy SM88 at the speed of movement of the mold: a – 5 mm/min.; b – 10 mm/min.

After heat treatment, dissolution of the γevt phase is observed in the microstructure of the test samples, which has a positive effect on the phase-structural stability of the castings.

The results of short-term strength tests (at V = 5 mm/min σB900°C - 646 MPa, δ = 16-18%; at V = 10 mm/min σB900°C - 650 MPa, δ = 18-20%) showed that the level of the obtained mechanical properties of all samples meets the requirements of the technical documentation for this product ($\sigma_B 900°C \leq 640$ MPa, $\delta \leq 9\%$).

The study of long-term strength was carried out at an operating temperature of 900°C, the load was 320 MPa. According to the test results, an increase in indicators is observed from τ = 110-115 h. at V = 5 mm/min to τ = 160-180 h. at V=10 mm/min), which confirms the high phase-structural stability of samples melted at V=10 mm/min. The level of long-term strength meets the requirements of the standard ≥ 100 hours (Instruction, 2011).

4 CONCLUSION

1. The technological process of the production of paddles has been improved due to the use of shell molds on a water binder with the use of polymer models.
2. It was determined that to obtain a regular directional structure, the speed of movement of the crystallization front is 10 mm/min. At the same time, the temperature gradient at the crystallization front is 165–175 °C/cm.
3. Tests of reference samples for short-term and long-term strength after heat treatment according to standard (Instruction, 2011) showed that the level of mechanical characteristics corresponds to that specified in the technical documentation for this product.

REFERENCES

Balitskii, A.I., Kvasnytska, Y.H., Ivaskevich, L.M., Kvasnytska, K.H., Balitskii, O.A., Miskiewicz, R. M., Noha, V.O., Parkhomchuk, Zh.V., Veis, V.I. & Dowejko, J.M. 2024. Improvement of hydrogen-resistant gas turbine engine blades: Single-crystal superalloy manufacturing technology. *Materials*, 17(17): 4265

Dexin, M.A. 2018. Novel casting processes for single-crystal turbine blades of superalloys. *Frontiers of Mechanical Engineering*, 13: 3–16.

Glotka, A.A., Haiduk, S.V. & Ol'shanetskii, V.Y. 2021. Modeling thermophysical characteristics of nickel-based superalloys. *Journal of Engineering Physics and Thermophysics*, 94: 1363–1368.

Instruction. Quality system. Technical requirements. Control and acceptance rules*: TI NPCG-157-2011*: *Introduced at SE GRPC "Zorya" - "Mashproekt"*, Nikolaev: 20.

Ivaskevich, L., Balitskii, A., Kvasnytska, I., Kvasnytska, K. & Myalnitsa, H. 2021. Thermal stability, cyclic durability, and hydrogen resistance of cast nickel-cobalt alloys for gas turbine blades. In *CAMPE 2021: Advances in Mechanical and Power Engineering*, 147–155.

Wisniewski, P., Sitek, R., Towarek, A., Choinska, E., Moszczynska, D. & Mizera, J. 2020. Molding binder influence on the porosity and gas permeability of ceramic casting molds. *Materials*, 13(12): 2735.

Welding and Related Technologies – Krivtsun et al. (Eds)
© 2025 The Author(s), ISBN 978-1-032-85176-1
Open Access: www.taylorfrancis.com, CC BY-NC-ND 4.0 license

Production of heat-resistant titanium based alloys by electron beam melting in Ukraine

S.V. Akhonin, A. Severyn & V.O. Berezos
E.O. Paton Electric Welding Institute, N.A.S. of Ukraine

V. Kryzhanovskyi
LLC Scientific and production association HVYLYA

M. Kuzmenko
I.M. Frantsevich Institute for Problems in Materials Science, N.A.S. of Ukraine

ABSTRACT: The article is devoted to the production of heat-resistant titanium alloys by electron beam melting in the conditions of import substitution of the titanium products market in Ukraine. The production of ingots of industrial heat-resistant titanium alloys and new generation heat-resistant alloys with a silicide hardening mechanism is described. It has been shown that the quality and properties of material and deformed semi-finished products are at the same level, and in some cases even exceed world analogues in terms of properties.

1 INTRODUCTION

The full-scale war, which was insidiously launched by the Russian Federation against Ukraine, increased the importance of providing the country with products of its own production, in particular those of strategic importance - both military and civilian. One of the strategic directions of industrial recovery should be the development of titanium production and industries based on it. This is due to a number of factors. Ukraine is one of the ten countries in the world that has both titanium ore reserves and a complete technological cycle of creating aviation equipment (Web portal of data on the mining industry of Ukraine, Shershakov & Shemaiev 2020). The development of the market and the application of titanium alloys is growing every day. As it is in the example of AIRBUS aircraft production. In the A350 model, the mass fraction of titanium alloys is 13% of the weight of the aircraft, as in the A320 model it was only 3%. Moreover, the majority of this share (~40%) is used in aircraft engines (Molyar et al. 2015). Increasing the competitiveness of aircraft engines primarily depends on their technical characteristics, as well as a significant reduction in the cost of production and operation. Part of the solution to these problems can be the use of fundamentally new structural materials, as well as high-tech methods of their production and processing, in the aircraft engines being developed.

When casting ingots of titanium alloys using traditional metallurgical technologies, namely vacuum arc remelting (VAR), in the process of casting, a significant overheating of the melt occurs, a large metal bath is formed, and as a result, liquation and a coarse-grained structure of the metal (Babenko & Dolzhenkova 2014). In recent years, the technology of electron beam melting (EBM) with intermediate power has significantly expanded the range of titanium-based alloys due to technological innovations, and is currently almost as good as the production technology of vacuum arc melting, traditionally used for the production of titanium ingots abroad. Thanks to the creation of a shallow liquid bath and, as a result, a finer and

DOI: 10.1201/9781003518518-8

more uniform structure, electron beam melting allows not only to clean the metal from gaseous and non-metallic impurities more deeply, but also significantly simplifies the process of metallurgical redistribution and ensures the production of products with qualitatively new physico-chemical and mechanical properties (Markovsky et al. 2020).

2 PRODUCTION OF HEAT-RESISTANT TITANIUM BASED ALLOYS BY ELECTRON BEAM MELTING

2.1 *Production of ingots of industrial heat-resistant alloys*

At present, heat-resistant industrial titanium alloys VT3-1, VT6 and VT20 are among the main materials used in low-pressure compressors of domestic engines of public and military aircraft, as well as in power engineering. These alloys were developed back in the 60s of the last century, but they are quite actively used (Il'in et al. 2009).

In order to study the quality of the metal and improve the productivity of the process, the comprehensive research work was carried out on improving the EBM technology of heat-resistant titanium alloys in the E.O. Paton Electric Welding Institute of the National Academy of Sciences of Ukraine (PWI) (Paton et al. 2010, Akhonin et al. 2019). When determining the technological parameters of the melts, the fundamental studies of the processes of evaporation of alloy components from the melt in a vacuum (Akhonin et al. 2003) and heating regimes of the ingot surface in the crystallizer, determined by mathematical modeling methods, conducted in PWI were used (Zhuk et al. 2005). Smelting of ingots was carried out in an industrial electron beam installation UE-5810 (Akhonin et al. 2022). Industrial batches of large ingots of heat-resistant titanium alloys VT3-1 with a diameter of 840 mm (Figure 1a) and VT6 with a diameter of 600 mm (Figure 1b) and VT20 with a diameter of 500 mm (Figure 1c) were obtained.

To assess the metal quality of the obtained ingots, a study of the chemical composition of samples taken along the length of the ingot was carried out. The results of the analysis of the chemical composition of the metal showed that the distribution of alloying elements along the length of the ingots is uniform and fully corresponds to the brand alloy composition. Deformed semi-finished products were made from the received ingots at Ukrainian enterprises.

A study of the mechanical properties and structure (Figure 1 a, b, c) of the obtained forgings made from EBM ingots was conducted (Table 1). The structure of the metal formed in the process of hot deformation significantly affects the mechanical properties of the products. The main factors here are the initial structure of the ingot, temperature, degree and speed of its deformation. Analysis of the microstructure of forged metal of the VT3-1 alloy showed that the microstructure of the metal samples corresponds to type 5-6 on the 9-type scale of microstructures. The microstructure of the forging of the VT6 alloy corresponds to type 3-4, and the microstructure of the deformed bars of the VT20 alloy corresponds to type 6-7 according to the 9-type scale of microstructures. All types of obtained microstructures meet the requirements of the standards.

The conducted experiments show that electron beam melting is an effective method of obtaining high-quality ingots of industrial heat-resistant titanium alloys and VT3-1, VT6 and VT20. EBM can completely replace the VAR process, and the quality and mechanical properties of deformed semi-finished products made from them fully meet the requirements of the standards.

2.2 *Production of ingots of heat-resistant titanium alloys of a new generation with a silicide hardening mechanism*

One of the directions for improving the high-temperature properties of titanium alloys is the creation of titanium composites based on the Ti-Si-Al system with silicon content. Such composites have a multiphase structure, where the strengthening of the titanium matrix by the refractory compound Ti_5Si_3 occurs naturally during the crystallization process. Alloys of this system are developed by the Institute for Problems in Materials Science (IPMS) of the National Academy of Sciences of Ukraine. These alloys are promising for creating a new class

Figure 1. Large ingots of industrial heat-resistant alloys VT3-1 (a) and VT6 (b) and VT20 (c) obtained by the EBM method and deformed semi-finished products made from them, and their structure.

Table 1. Mechanical properties of forgings made from EBM ingots of industrial heat-resistant titanium alloys.

	The average indicator of mechanical characteristics from a batch of samples				
Alloy	UTS, MPa	δ, %	ψ,%	KCU, Dj/sм2	Hardness, HB
VT3-1	1152	11	24	36	352
Requirements according to the standard	950-1200	≥9	≥22	≥30	269-363
VT6	960	12	42	12	280
Requirements according to the standard	950-1150	≥10	≥30	≥4	269-363
VT20	942	11	23	5	-
Requirements according to the standard	900-1150	≥18	≥20	≥3	255-341

of materials with a high level of heat-resistant characteristics and due to their relatively low cost (Firstov et al. 2009, Firstov, et al. 2019, Kuang et al. 2017, Shevchenko et al. 2023).

Ingots Ø110 mm of heat-resistant alloy system Ti-(6.2-6.9)Al-(5.0-5.5)Zr-(0.5-0.85)Si-(0.5-0.8)Mo-(0.5-0.8)Nb-(1.5-2.5)Sn (wt.%) were obtained at the electron-beam furnace (Akhonin et al. 2022). The metal structure of the ingots is equiaxed, dense, homogeneous, with no zones that are etched differently across the cross section of the ingot. There is no significant difference in the structure of the central zone of the casting and that of the peripheral zone (Figure 2a). The macrostructure of the longitudinal template shows a small liquid bath during the entire melting process (Figure 2b). Defects in the form of pores, shells, cracks and non-metallic inclusions were not detected. In none of the obtained ingots, the segregation of alloying elements, characteristic of VAR ingots, was detected. The study of the structure of the obtained alloys showed that they can be attributed to pseudo-α-alloys: they consist of packets of lamellae of lamellar α-phase and a small amount of residual β-phase. Silicides are also present in the structure in the form of both whole layers and small secondary silicides along the boundaries of α-plates (Figure 2c, 2d).

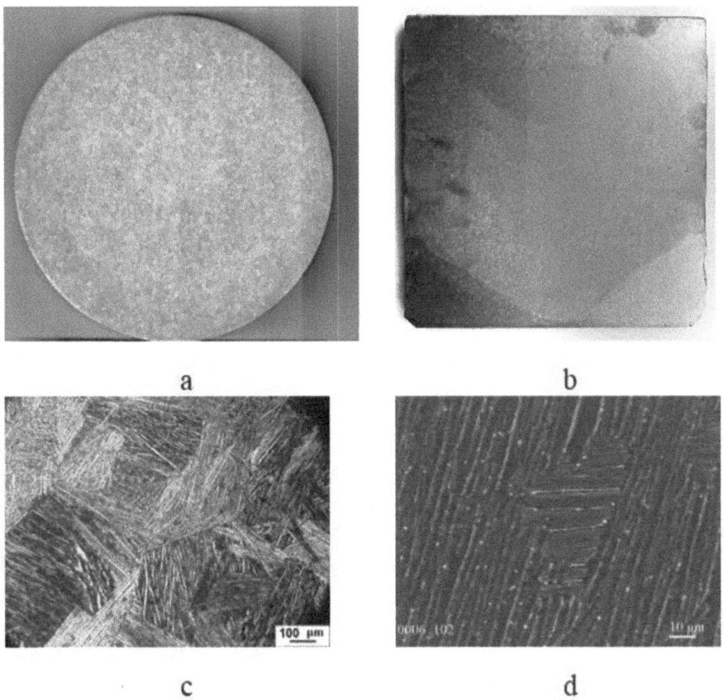

Figure 2. Macrostructure of the cast metal of a Ø110 mm ingot of the Ti-Al-Zr-Sn-Mo-Nb-Si system alloy (a – transverse template; b – longitudinal template) and its microstructure (a - optical microscopy b - raster electron microscopy).

Work was carried out on deformation processing of EBM ingots (Fig. 4a) of experimental alloys and deformed semi-finished products with a diameter of 55 mm and a length of 1200 mm were obtained (Fig. 4b). Deformation treatment of EBM ingots of the Ti-Al-Zr-Sn-Mo-Nb-Si system forms a mixed globular-plate structure, where the α-phase is present in both plate and globular form (Fig. 4b). The average grain size is 20-40 μm. A large number of silicides is observed in the structure, but they are quite dispersed and distributed between α-plates and in the form of separate allocations along the boundaries and in the middle of α-phase grains. X-ray structural analysis studies confirmed that the pseudo-β heat-resistant titanium alloy is a two-phase alloy and consists of the beta phase of titanium (β–Ti) in the amount of 2.4% and the alpha phase of titanium (α–Ti) in the amount of 97.6 %.

A study of the mechanical properties of cast and deformed metal alloy of the Ti-Al-Zr-Sn-Mo-Nb-Si system at room temperature and at temperature of 600°C was carried out (Table 2).

a b

Figure 3. Deformed semi-finished products of the heat-resistant alloy of the system Ti-Al-Zr-Sn-Mo-Nb-Si Ø55 mm (a) and its microstructure (b).

Table 2. Tensile mechanical properties of the cast and deformed alloy at different temperatures.

	Test temperature, °C					
	20°C			600°C		
Metal condition	UTS, MPa	YS, MPa	δ,%	UTS, MPa	YS, MPa	δ,%
Cast	1056	1033	0,38	674	570	4
Deformed	1160	1102	9	738	621	18
IMI834*	1040	932	12	600	500	15

* For comparison (TMC-0164. 2000)

Studies show that the plasticity of the material at room temperature, compared to cast, increases almost 25 times, and the high strength indicators both at room and at working temperatures of 600°C exceed the level of the best world analogues, which indicates the potential possibility of using these alloys in aircraft engines of domestic production.

Technological processes of EBM, developed in PWI makes it possible to obtain ingots of heat-resistant titanium alloys that meet all the requirements for chemical composition and mechanical properties. Installed in SE "RPC "Titan" E.O. Paton Electric Welding Institute of the National Academy of Sciences of Ukraine equipment allows to produce industrial batches of titanium ingots of various standard sizes, to expand the scope of application of titanium in various industries and to organize competitive production of ingots of heat-resistant titanium alloys in Ukraine.

REFERENCES

Web portal of data on the mining industry of Ukraine. Web link https://eiti.gov.ua/resursi-rozvidka-ta-vidobuvannya/rudi-titanu_2022/ (in Ukrainian).

Shershakov O.M. & Shemaiev V.M. 2020. Regarding the implementation of the strategy for the revival of the domestic aircraft industry for the period up to 2030. *National Institute of Strategic Studies. Research. National security. Web link* https://niss.gov.ua/sites/default/files/2020-12/az-09.12.20_1-1.pdf (in Ukrainian).

Molyar A.G. et al. 2015. *Structural materials in aircraft construction.* Kyiv: KVITS. -320 (in Russian).

Babenko E.P. & Dolzhenkova E.V. 2014. Study of the causes of destruction of a large-sized product made of VT23 alloy. *Metallurgicheskaya i gornorudnaya promyshlennost'.* #3, p.82–85 (in Russian).

Markovsky P.E. et al. 2020. Microstructure and Tensile Properties of Cost-Efficient Thermally Hardenable α + β Alloys of Ti-Al-Mo-Fe and Ti-Al-Mo-Cr Systems. Metallography, Microstructure, and Analysis. 9(6), pp. 856–872.

Il'in A.A. et al. 2009. *Titanium alloys. Composition, structure, properties: Handbook*. Moscow: VILS-MATI -520. (in Russian).
PatonB.E. et al. 2010. Production of large-sized ingots of heat-resistant titanium-based alloys by electron beam melting. *Sovremennaya elektrometallurgiya*. №3. p. 11–14 (in Russian).
Akhonin S.V. et al. 2019. Structure and Properties of Titanium Modified Silicon-carbide at EBM. *IOP Conference Series: Materials Science and Engineering*, # 582 (1).
Akhonin S.V. et al. 2003. Mathematical modeling of aluminum evaporation during electron-beam cold-hearth melting of Ti-6Al-4V ingots. *Metall Mater Trans*. B 34, p.447–454.
ZhukG.V et al. 2005. Mathematical modeling of thermal processes in EBCHM of hollow ingots. *Problemy Spetsial'noj Electrometallugii*. Issue 4, P. 16–18.
Akhonin S. et al. 2022. Determining the structure and properties of heat-resistant titanium alloys VT3-1 and VT9 obtained by electron-beam melting. *Eastern-European Journal of Enterprise Technologies*, 5 (12 (119), p.6–12.
Firstov S.A. et al. 2009. Titanium "irons" and titanium "steels". *Metal Science and Heat Treatment*. Volume 51, Issue 1-2, P. 12–18.
Firstov, S.O. et al. 2019. Alloys of the Ti–Al–Zr–Si System Intended for Operation at High Temperatures. *Materials Science*. Volume 54, Issue 6, 1 May, P. 783–788.
Kuang Q.B. et al. 2017. Microstructure and Properties of High Temperature Titanium Alloys with a High Si. Content Prepared by Powder Metallurgy. *Materials Transactions*. Vol. 58, No. 12. p. 1735–1741.
Shevchenko O.M. et al. 2023. The influence of the deformation and heat treatment on the structure and heat-resistance of Ti-Al-Zr-Si alloys. *Materials Science*. V59, No. 1. p. 40–48.
Akhonin S.V. et al. 2022. Producing high-temperature titanium alloys of Ti-Al-Zr-Si-Mo-Nb-Sn system by electron beam melting. *The Paton Welding J*., 7, p.39–45.
TMC-0164. 2000. Timet DataSheet Alloy 834. *Timet*. 3p.

Welding and Related Technologies – Krivtsun et al. (Eds)

Aluminothermic reduction of vanadium-containing waste under conditions of electroslag smelting

M.O. Vdovin
E.O. Paton Electric Welding Institute (PWI) Kiev, Ukraine
PlasmaTec LLC, Ukraine

Yu.V. Kostetskyi, Ye.O. Pedchenko, G.O. Polishko, V.P. Petrenko & V.A. Zaitsev
E.O. Paton Electric Welding Institute (PWI) Kiev, Ukraine

ABSTRACT: The paper presents the results of studies carried out under pilot scale conditions on the production of ferrovanadium by the electro-aluminothermic method from chemically enriched vanadium-containing wastes with reduction of oxides in the slag layer. Carrying out the reduction process in the slag layer allows the intensity of the process to be controlled and produces a ferroalloy with a vanadium content of 40 to 80%. The electroslag process allows better separation of the smelting products and has significant potential for using various man-made wastes in the feed to reduce costs. During the research, crucibles with different linings were tested and the most efficient electrical modes were determined. The power source for the melting unit was a DC welding transformer. It was carried out experiments with different polarity of the power supply. As a result, a technology was adapted to produce a ferroalloy with a vanadium content of 50-60% based on the use of technogenic waste for electrode production. In pilot melts the recovery of vanadium in metal was 85-90%. The technology ensures high product quality and the possibility of waste-free production. The investment component in equipment for this process is relatively low.

Keywords: ferrovanadium, electroslag process, aluminothermic reduction, man-made waste

1 INTRODUCTION

The main consumer of vanadium is the metallurgical industry, which consumes about 90% of all vanadium mined in the world for alloying steel and alloys, mainly in the form of ferrovanadium (Vanadium 2021, Simandl 2022). The rest of vanadium is used in the form of V_2O_5, for the manufacture of catalysts, and other compounds for medical and other purposes. Today, the ferroalloy market offers ferrovanadium grades with a vanadium content of 35-85% (Gasik 2013, Gasik 2021). However, the largest consumption is accounted for by the FeV80 brand (Vanadium 2021).

In light of the growing demand for ferrovanadium, particularly in Ukraine, and the increasing amount of vanadium-containing man-made waste, it is prudent to explore the potential of using these materials as a raw material for vanadium extraction (Simandl 2022). Usually, the raw materials for smelting ferrovanadium are its oxides (Vanadium 2021, Gasik 2013). Vanadium forms several oxides. These are VO, V_2O_3, VO_2, V_2O_5 and some intermediate forms. The most interesting for recovery are V_2O_3 and V_2O_5 (Swinbourne 2016). In man-made waste, vanadium oxides are present in the products of fuel oil combustion, alumina production, titanium and magnesium production and in spent vanadium catalysts and batteries. This kind of waste is relatively easy to chemically enrich from 3-5% vanadium oxide to 80-85%

DOI: 10.1201/9781003518518-9

(Wenfang 2020). An example is the resulting concentrate with the following phase composition of vanadium oxide: V_2O_5 55-65%, V_2O_3 15-20%, VO_2 3-7%, VO 0.5-4%.

Silicon, hydrogen, carbon and aluminum can be used to recover vanadium from oxides (Gasik 2021, Swinbourne 2016). In industrial processes, for obvious reasons, aluminum is usually used (Gasik 2013, Lyakishev 1978). The reactions of reduction of vanadium oxides with aluminum are exothermic, which significantly reduces the cost of electrical energy and facilitates the organization of the reduction process. At the same time, a certain aluminum content in the final product does not interfere with the alloying of steels. Due to the possibility of adding additional heat, the electroaluminothermic process makes it possible to use lower vanadium oxides in the charge, the reduction reactions of which with aluminum are less exothermic than for V_2O_5 (Swinbourne 2016). Moreover, the electroslag process allows for better separation of smelting products and control of the intensity of the process. All other reduction options are endothermic processes, and some of the reducing agents are undesirable impurities (Gasik 2021).

The electroslag process has significant potential for using various man-made wastes in the feed to reduce costs. For example, aluminum waste is readily available at relatively low cost and the enriched alumina slag, a by-product of the reaction, is recycled. To regulate the vanadium content in the alloy, it can use small steel waste after metalworking (shavings, chips, shot, etc.). If desired, steel powder can be replaced with scale, the waste of which is also generated a lot in rolling or drawing industries. In this case, of course, it will be necessary to increase the consumption of the reducing agent.

In this work, the industrial process of electroaluminothermic reduction of vanadium from a technical vanadium pentoxide concentrate under electroslag smelting conditions to obtain ferrovanadium with a given vanadium content has been studied. These pilot meltings have been carried out to test the technology to produce ferrovanadium with 50-60% vanadium for electrode production.

2 EXPERIMENTAL METHODOLOGY

Experimental smelting was carried out on a modified electroslag flux melting unit of the OB-1901 type. Its column with a non-consumable electrode has a rotary mechanism and can move the electrode from the axis of the melting space by 90°, and the crucible can turn in the vertical plane by 180°. The power source of the smelting plant is a welding transformer of direct current at 1 kA and a voltage of 50 V. The installation has an automatic voltage control and a clear adjustment of current and voltage at the power supply itself. The melting of the charge takes place in a crucible lined with refractory material.

During the experiments, the charge for the electroslag reduction process consisted of vanadium pentoxide concentrate, secondary aluminum, waste steel powder, and fluxes to reduce the refractoriness of the slag. Calcium oxide is added to the charge to bind alumina into less refractory calcium aluminate (Hallstedl 1990).

All charge materials with a fraction of 0.3-1.0 mm were mixed in a mixing drum and sent to the melting unit. The charge mixture was loaded into the gap between the graphite electrode and the crucible wall. The thermite reaction was initiated by an arc, and then, as the charge melted, a new portion of the charge was added to the crucible. After the entire charge was melted, the slag bath was kept under current for about 10 minutes.

The reduction of vanadium with aluminum takes place in the slag layer. During the smelting process, metal droplets settle to the bottom of the crucible under a protective layer of slag. The electroslag process ensures smooth recovery and a liquid state of the growing slag layer, both by the energy of the thermite reaction and by electrical energy. Based on the experience of previous studies (Kostetskyi 2023), to ensure an acceptable level of vanadium extraction, the reducing agent was added to the charge in an amount of 101% to 103% of the stoichiometrically calculated amount. Further increases in the amount of reducing agent will only worsen the melting process and increase the aluminum content of the final product.

3 DISCUSSION OF RESULTS

The first series of experimental melts (Nos. 1-4, Table 1) were carried out with a power supply connection according to the scheme - "minus" on the main electrode, and "plus" on the bottom electrode. The crucible was lined with Al_2O_3, and the bottom electrode was made of graphite During the experimental tests, the parameters of the technological process were adjusted in order to ensure the desired course of the process and to obtain an alloy of the target composition. The composition of the charge of the experimental melting is given in the Table 1. The chemical composition of the resulting metal of the first series of melts is shown in the Table 2.

As can be seen, the alloy has an increased carbon content because of its interaction with the bottom electrode material. In addition, the experimental alumina-based lining showed low stability under process conditions. Therefore, it was decided to re-design the bottom electrode and replace the lining.

Table 1. The composition of the charge (%).

Melt	V_2O_5	Al	Fe	CaO	CaF_2
1, 2	46.3	22.5	7.6	19.3	4.2
3, 4, 5, 6	46.27	19.2	7.4	24.4	2.5
7, 8, 9	48.65	20.3	7.8	20.3	2.9
10, 11, 12	46.75	22.1	7.5	20.7	2.85

Table 2. The chemical composition of the metal in the first series of melts (%).

Melt	V	Al	Si	Ti	Mn	Fe	C
1, 2	55	7.1	2.2	0.3	0.68	32.5	1.15
3, 4	65	1.4	1.0	0.2	0.50	29.3	1.40

The next series of melts (Nos. 5-12, Table 1) were conducted in a modernized crucible with a refractory magnesite brick lining and a steel bottom electrode. The polarity of the power supply was also reversed.

The chemical composition of the metal obtained during melting in the modernized crucible is given in the Table 3. The change in the design of the bottom electrode ensured the production of an alloy with the proper carbon content. The results of the last three melts (Nos. 10-12, Table 1) have shown that the defined parameters of the technology ensure the stable production of an alloy with the target chemical composition. Each of these melts yielded an average of 7.15 kg of ferrovanadium, which is 50.35% of the total vanadium oxide and iron in the charge, and 15.9 kg of slag. Losses with fumes and evaporation were about 0.55 kg. The extraction of vanadium into the metal was 85-90%. Based on the results of experimental melting, the electric mode and other parameters of the technology were determined, which ensured the stable production of ferroalloy grades FeV50, and FeV60 (Nos. 10-12, Table 3).

Table 3. The chemical composition of the metal in the second series of melts (%).

Melt	V	Al	Si	Cr	Mn	Fe	C
5	59.99	2.98	0.75	0.68	1.11	33.24	0.59
6	63.09	2.16	0.78	0.55	2.54	29.71	0.42
7, 8, 9	56.60	2.30	1.5	3.40	2.90	32.10	0.30
10, 11, 12	68.70	1.80	1.0	0.20	0.09	27.80	0.30

4 CONCLUSIONS

Pilot scale studies on the production of ferrovanadium by the electro-alumina process from chemically enriched vanadium-containing wastes using the electroslag process have shown that the reduction of vanadium oxides in the slag layer, with the possibility of supplying additional energy, allows effective control of the reduction process intensity.

The electroslag process ensures efficient separation of the smelting products and has significant potential for using various industrial wastes in the charge to reduce costs. Experiments have shown that by adjusting the composition of the charge it is possible to produce a ferroalloy with a vanadium content of 40 to 80%.

It has been found that when using a DC power supply, it is recommended to connect the power supply so that the main electrode has a 'plus'.

During the test melts, a technology was adapted to produce a ferroalloy with a vanadium content of 50-60% based on the use of technogenic waste for electrode production. In the pilot melts the recovery of vanadium in metal was 85-90%. This indicator can be improved by further improving the technology.

The electroslag reduction process ensures high product quality and the possibility of waste-free production. The investment in equipment for this process is relatively low.

REFERENCES

Vanadium: Extraction, Manufacturing and Application. (2021) Elsevier. DOI: 10.1016/C2018-0-03399-X

Simandl G. J. (2022) Vanadium as a critical material: economic geology with emphasis on market and the main deposit types. Applied Earth Science. 131 (4). 218–236. DOI: 10.1080/25726838.2022.2102883

Gasik M. (2013) Technology of vanadium ferroalloys. Handbook of Ferroalloys. Elsevier. pp. 397–409

Gasik M., Dashevskii V., Bizhanov A. (2021) Ferroalloys: Theory and Practice. Springer.

Wenfang G., Zhi S., Hongbin C., He D., Yujiao Z., Pengge N., Gaojie X., Yi Z. (2020) Economic evaluation of typical metal production process: A case study of vanadium oxide production in China. Journal of Cleaner Production. 256. 120217. DOI: 10.1016/j.jclepro.2020.120217

Swinbourne D. R., Richardson T. Cabalteja F. (2016) Understanding ferrovanadium smelting through computational thermodynamics modelling. Mineral Processing and Extractive Metallurgy. 125 (1). pp. 45–55. DOI: 10.1179/1743285515Y.0000000019.

Lyakishev N.P., Pliner Yu.L (1978) Aluminothermia. Moscow: Metallurgy.

Hallstedl B. (1990) Assessment of the CaO-Al2O3 System. Journal of the American Ceramic Society. 73 (1). pp. 15–23. DOI: 10.1111/j.1151-2916.1990.tb05083.x

Kostetskyi Yu. at al. (2023) Features of the Process of Obtaining Ferrovanadium under Conditions of Electroslag Melting. Proceedings 32nd International Conference on Metallurgy and Materials METAL 2023. Brno. Czech Republic. pp. 69–76. DOI: 10.37904/metal.2023.4629.

Session 3. Innovative technologies, materials and equipment for welding and related processes

Welding and Related Technologies – Krivtsun et al. (Eds)

Using spectral parameters of compressed arc plasma to develop basic approaches to smart welding

V.M. Korzhyk, S. Gao & X. Wang
China-Ukraine Institute of Welding, Guangdong Academy of Sciences, Guangdong Provincial Key Laboratory of Material Joining and Advanced Manufacturing, Guangzhou, China

V.Ya. Chernyak
Taras Shevchenko National University, Kyiv, Ukraine

V.Yu. Khaskin, I.M. Klochkov, Ye.V. Illiashenko, O.M. Voitenko, A.A. Grynyuk & O.V. Konoreva
E.O. Paton Electric Welding Institute of the NAS of Ukraine, Kyiv, Ukraine

ABSTRACT: It is proposed to analyze the spectral composition of the arc plasma on-line by pre-selected lines using CCD detectors with a sensitivity range of 190÷1100 nm and a response time of 4 ms. The analysis performed will allow recording the overall temperature (energy) of the plasma and the stability of welding. The information obtained will be analyzed by self-learning neural networks to improve the quality of automated control of welding processes.

1 INTRODUCTION

One of the modern progressive trends of investigation and diagnostics of welding processes is application of on-line plasma optical spectroscopy (Valdiande 2022, Han et al. 2022, Gu et al. 2022). Measurements of such parameters of the welding arc as concentration and distribution of iron vapours, plasma temperature, and electron density allow controlling the state of local thermodynamic equilibrium of plasma (Valensi et al. 2010). It provides information about the stability of the welding process, possibility of weld defect formation, etc.

However, there exist a number of problems associated with on-line processing of the results of welding arc plasma spectroscopy and obtaining the necessary information. It is necessary to clearly determine the least noisy spectral lines of plasma radiation, which will be followed during welding. More over, it is necessary to have certain standards, with which the followed lines will be compared. Such standards should be in place both for the cases of sound weld formation and for the cases of formation of various defects. Recording of plasma spectra and their comparative identification should take place automatically at sufficiently high speeds (for instance, close to weld pool self-oscillations (Lebedev & Novykov 2022)).

2 OBJECTIVE

The objective of the work is creation of the concept of application of dynamic spectral analysis of arc plasma in intelligent welding processes and related technologies to improve the quality of on-line diagnostics of weld formation and further development of the system of automatic monitoring of the processes.

DOI: 10.1201/9781003518518-10

To achieve the defined objective, the following concept of the procedure for further research performance is proposed. First, it is necessary to select the clearest spectral lines (the least noisy, accordingly) of certain frequencies, characteristic for welding a certain metal or alloy. Application of spectral lines of plasma or shielding gases used in the welding process is also possible. Then it is necessary to obtain the reference spectra of the selected frequencies for cases of sound formation of the weld or deposited metal bead in the welding process. After that, at realization of arc or plasma-arc processes of welding and surfacing it will be possible to create a system of spectral analysis of welding arc plasma, which will perform continuous comparison of the current spectra of the selected frequency lines with the reference ones. Obtained results will allow drawing conclusions about the behaviour of the weld pool and quality of formation of the weld or deposited metal bead.

3 RESEARCH AND DISCUSSION

It is rational to take measurements of arc plasma spectrum by spectrometers with CCD-detectors, with a broad sensitivity range (190÷1100 nm) and rather high response (from 4 ms). This allows analyzing the spectral composition of welding arc plasma by pre-selected lines (Lim et al 2021). Such analysis should be dynamic, i.e it should be performed directly during welding. It can be built into the welding control system through the respective feedbacks. We can also limit ourselves to recording the deviations of the stability of the process of determination of the zones of postweld monitoring of the weld.

As was noted above, spectral analysis of the welding arc allows recording the concentration and distribution of iron vapours, plasma temperature and electron density, and local thermodynamic equilibrium of plasma. The authors propose monitoring the overall temperature (energy) of arc plasma, population of electron levels of the respective spectra, charge concentration and degree of plasma ionization. This will allow realization of a system of monitoring and control of the welding processes, based on spectral analysis of the welding arc (Figure1). As a result, it will be possible to analyze the energy input into the weld pool, depending on external destabilizing factors (for instance, instability of power source operation or defects of assembly of the butt being welded). More over, it will allow eliminating the risk of lacks-of-penetration, blowholes and burns-through in the welds.

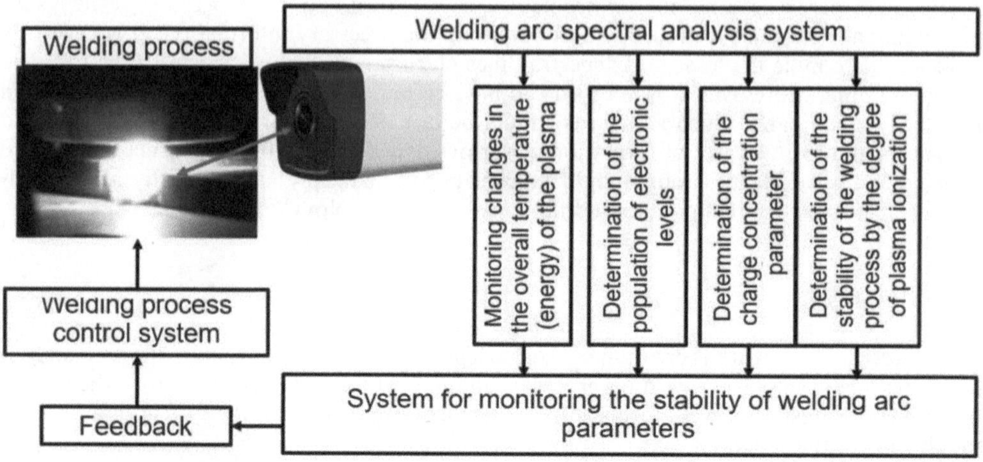

Figure 1. Scheme of the system of monitoring and control of the welding processes based on spectral analysis of the welding arc.

It is planned that further use of the obtained results will promote creation of a new approach to the process of development of intelligent welding. For realization of such a process the authors propose introducing a feedback into the welding system, which follows the spectral composition of the welding arc plasma (Figure2). System of diagnostics of the spectral composition of welding arc plasma can be one of the elements of the general system of diagnostics and monitoring of welding process parameters on-line (remote access). Such a general self-learning system of diagnostics and monitoring using an artificial neural network can perform automatic intelligent processing and adjustment of the welding process parameters (Lebedev & Novykov 2022).

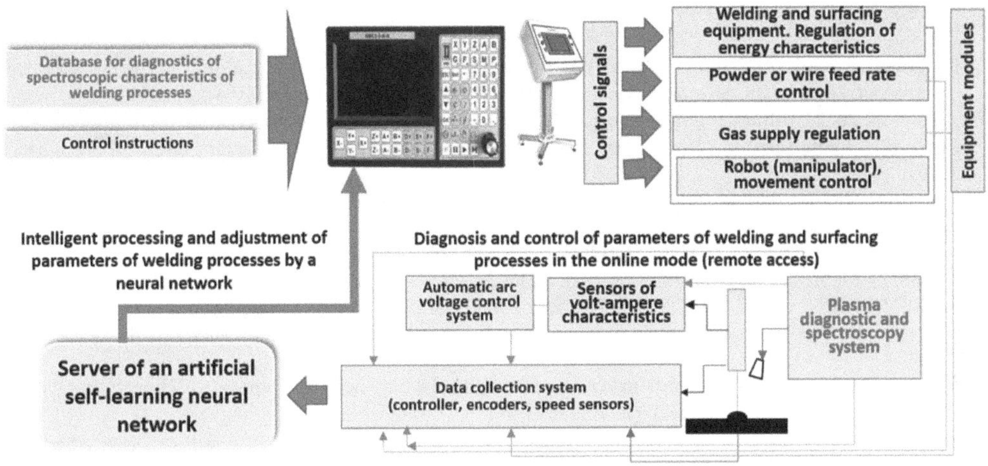

Figure 2. Scheme of an intelligent system of welding process control with feedback, including spectrometric diagnostics.

Another approach to application of welding arc spectral analysis is use of neural networks, which will allow performance of welding equipment self-learning, and replacing the welding operator as much as possible in the future. The recorded information about the spectral composition of the welding arc and its behavior will be analyzed by the neural networks. This information will be the base of feedback system operation. As a result, the neural networks will perform high-speed comparison of standard spectral lines with those followed directly during welding. This will enable performance of on-line monitoring of the quality of weld formation, recording defect development and their elimination to some extent. The obtained result will promote improvement of control of the welding processes (as in (Kvasnytskiy et al. 2022, Korzhyk et al. 2021, Sydorets et al. 2017)) and of the derived results.

4 CONCLUSIONS

Thus, a new approach is proposed for development of intelligent plasma-arc welding process, which includes presence of feedback in the technological complex, which follows the welding arc spectral composition. Owing to application of neural networks with self-learning function, in a broad spectral range of sensitivity (190÷1100 nm) of CCD-detectors of plasma spectroscopes, and their rather fast response (from 4 ms), this will allow performing on-line analysis of the spectral composition of welding arc plasma by pre-selected spectral lines. The self-learning neural networks will be able to analyze the information on the welding arc spectral composition, its luminosity and visually recoded behaviour, and automate the welding process control.

ACKNOWLEDGMENTS

The work was funded within the following programs:

1. The National Key Research and Development Program of China [grant number 2020YFE0205300], China.
2. GDAS'Project of Science and Technology Development (2020GDASYL-20200301001), China.
3. Research projects of China [G2023030004L], China.

REFERENCES

Valdiande, J.J. et al. 2022. On-line monitoring and defect detection of arc-welding via plasma optical spectroscopy and LIBS. *Spectrochimica Acta Part B: Atomic Spectroscopy*. (194): 106474.

Han, J. et al. 2022. Minimizing defects and controlling the morphology of laser welded aluminum alloys using power modulation-based laser beam oscillation. *Journal of Manufacturing Processes*. (83): 49–59.

Gu, Y. et al. 2022. Corrosion resistance of 316 stainless steel in a simulated pressurized water reactor improved by laser cladding with chromium. *Surface and Coatings Technology*. (441): 12853.

Valensi, F. et al. 2010. Plasma diagnostic in gas metal arc welding by optical emission spectroscopy. *Journal of Physics D: Applied Physics*. (43): 434002.

Lebedev, V.O. & Novykov, S.V. 2022. Influence of weld pool low-frequency oscillations on the formation of crystallites size and welded metal microstructure. *Metallophysics and Advanced Technologies*. 44(3): 419–430.

Lim, J.S. et al. 2021. Measurement of electron density in transient spark discharge by simple interferometry. *Results in Physics*. (20): 103693.

Kvasnytskyi, V. et al. 2020. Designing brazing filler metal for heat-resistant alloys based on ni al intermetallide. *Eastern-European Journal of Enterprise Technologies*. 6(12): 6–19.

Korzhyk, V. et al. 2021. Comparing features in metallurgical interaction when applying different techniques of arc and plasma surfacing of steel wire on titanium. *Eastern-European Journal of Enterprise Technologies*. 4(12): 6–17.

Sydorets, V. et al. 2017. Electrical characteristics of the equipment for the hybrid plasma-MIG welding. *58th Annual International Scientific Confererence on Power and Electrical Engineering of Riga Technical University, RTUCON 2017 - Proceedings, 2017-November*: 1–6.

Welding and Related Technologies – Krivtsun et al. (Eds)

Brazing of titanium alloy TA15

S.V. Maksymova, V.V. Voronov & P.V. Kovalchuk
E.O. Paton Electric Welding Institute of NAS of Ukraine

ABSTRACT: This paper presents the results of research on the influence of the technological parameters of the brazing process on the structure formation of the titanium alloy TA15 brazed joints. The effects of surface preparation and brazing modes on the spread area of the Ti-Zr-Me filler metal on TA15 titanium alloy were analyzed. Findings indicate that the spread area for samples subjected to mechanical cleaning is 10-20% larger than for samples treated by standard acid etching (HNO_3+HF). Additionally, increasing the heating temperature from 900 °C to 950 °C enlarges the spread area by 10%. The study also explored the structuring of brazed joints, revealing that extending the exposure time during brazing increases joint width, homogenizes the structure, and boosts shear strength by 15%.

1 INTRODUCTION

Among the methods of joining titanium and its alloys, brazing processes occupy one of the leading places. This method makes it possible to create strong, hermetic, and heat-resistant connections, which is critical for many industries, such as aviation, energy, automotive, electronics, and electrical engineering (Gillespie 1999). At the same time, the technology of brazing and filler metal should provide brazed joints with properties close to the properties of the base material (Shapiro 2017). In general, modern methods of brazing titanium alloys are focused on improving the mechanical properties of joints by reducing the number of intermetallic compounds formed at the interface, both by optimizing the composition of filler metals and the technological parameters of the brazing process (Jing et al. 2016, Jing et al. 2017, Jing et al. 2020, Pang et al. 2016). At the same time, it is the temperature-time limitations of the parameters of the brazing cycle, due to the presence of the $\alpha \rightarrow \beta$-transformation, that are basic (Muhrat et al. 2018, Shapiro et al. 2007).

The TA15 alloy is a titanium pseudo-α alloy, due to its good high-temperature mechanical and technological characteristics up to temperatures of 500...600 °C, it is considered as one of the potential materials for the manufacture of gas turbine blade structures for aerospace purposes, etc. (Jing et al. 2016). However, it should be noted that not enough research has been conducted to study the brazing process of this material (Jing et al. 2020, Jing et al. 2020). Data available in the literature on the brazing of this material with filler metals of the TiZr-CuNi system indicate that the filler metals of this alloying system are able to provide good mechanical properties of brazed joints, both at room and at elevated (550...600 °C) temperatures (Jing et al. 2020a, Jing et al. 2020b, Huai et al. 2013). The strength of butt-brazed joints can exceed 90% of the strength of the base metal both at room and at elevated 550 °C (Huai et al. 2013), 600 °C (Jing et al. 2020a, Jing et al. 2020b) temperatures.

At the same time, brazed joints made using of this type filler metal have a heterophase structure consisting of a Ti-based solid solution matrix and a significant amount of brittle in-termetallic compounds Ti_2Cu, Ti_2Ni, Ti_3Cu_4 and δ-TiCu (Rabinkin et al. 1991, Shapiro 2017). It is indicated that with long-term holdings, it is possible to avoid the formation of

DOI: 10.1201/9781003518518-11

intermetal-lic compounds in the seam and to obtain a homogeneous Widmanstedt-type structure in the area of the brazed seam (Jing et al. 2020).

The purpose of this study was to investigate the impact of holding time during the brazing process of the TA15 titanium alloy on the structural formation and mechanical properties of the brazed joints.

2 MATERIALS AND METHODOLOGY

During the experiments, the TA15 titanium pseudo-α alloy was used as the base metal (Table 1).

Table 1. Chemical composition of TA15 (Ti-rest).

Fe	C	Si	Mo	V	Al	Zr	O	H
0.25	0.10	0.15	0.50 - 2.00	0.80 - 2.50	5.50 - 7.00	1.50 - 2.50	0.15	0.015

A Ti-Zr-Me (Cu, Ni)-based filler metal in the cast state was used for brazing. To evaluate the influence of surface preparation of the base metal on the spread of the filler metal, samples were cleaned either mechanically or through chemical etching in a mixture of HNO_3 and HF acids. Brazing was carried out in a vacuum furnace (SGV 2.4-2/15-I3) with radiation heating at a working space pressure of 1.33×10^{-3} Pa. The heating rate was 18-20 °C/min, and the cooling rate in the temperature range of 1000-200 °C was 10-15 °C/min, with an accuracy of temperature measurement of ±5 °C. To assess the effect of exposure time on the structure and properties of brazed joints, experiments were conducted with exposure times of 15, 30, and 60 minutes. Prior to brazing, the base metal blanks were processed according to standard methods.

Metallographic studies were conducted using a Tescan Mira 3 LMU scanning electron microscope. The distribution of elements in different phases was analyzed by local micro-X-ray spectral analysis using an Oxford Instruments X-max 80 mm^2 energy dispersive spectrometer. Sections were examined without chemical etching in backscattered electron (BSE) mode. The mechanical properties were tested at room temperature using a ZDM 10 Zwick-1488 testing machine.

3 RESULTS AND DISCUSSION

To evaluate the influence of the temperature on the spread of the filler metal on the TA15 base metal, a series of experiments was conducted at temperatures of 900 °C, 950 °C, and 1000 °C, each with a holding time of 15 minutes. The results of these studies show that as the heating temperature increases from 900 °C to 950 °C, the spread area of the filler metal significantly increases (Figure 1).

It should be noted that the surface preparation of the base metal significantly affects the spread area of the filler metal. The results indicate that for samples subjected to mechanical cleaning, the spread area is 10-20% greater than that of samples processed by standard etching in a mixture of acids (HNO_3 + HF) (Figure 1).

Metallographic studies of brazed joints revealed that at a brazing temperature of 900 °C with a holding time of 15 minutes (Figure 2a), a coarse slatted structure forms in the seam, characterized by wide, disoriented α-phase subgrains (dark) and broad interlayers (light) filled with acicular β-phase, alloyed with other components of the filler metal (Cu, Ni). This is evidenced by the increased concentration of these elements observed during electron beam scanning.

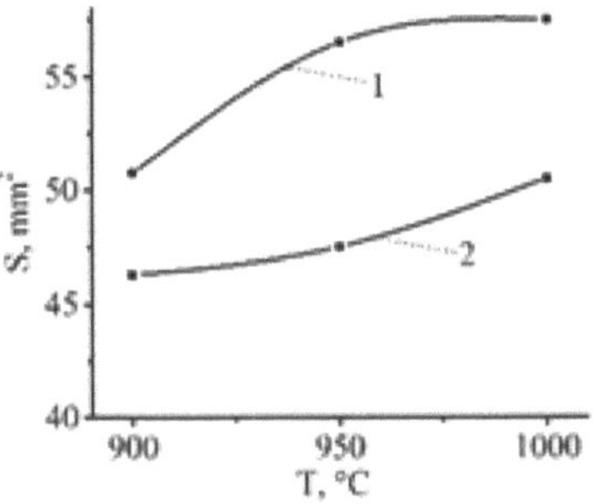

Figure 1. Spreading area Ti-Zr-Me brazing filler metal on mechanically cleaned (1) and etched (2) TA15 alloy samples.

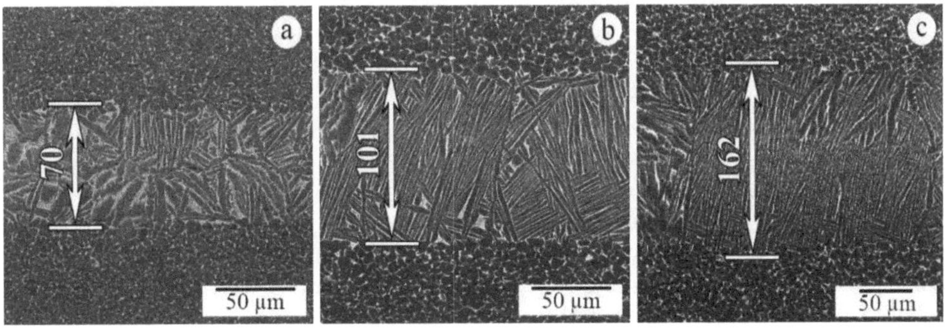

Figure 2. Microstructure of brazed joints obtained at different exposure times (min): 15 (a); 30 (b) and 60 (c).

As known from the literature, these elements lower the temperature of the $\alpha \rightarrow \beta$ transition, contributing to the formation of a martensite-like acicular phase. The width of the brazed joint is approximately 70-80 microns.

If the exposure time is increased to 30 minutes (Figure 2b), the seam width expands to 100 μm and the amount of α-phase (dark) increases. Simultaneously, the width of the acicular β-phase decreases (Figure 2b). Single peaks of copper and nickel in the interlayers of the acicular phase are observed on the element distribution line.

Further increasing the brazing duration to 60 minutes enhances the homogeneity of the structure of the brazed joint (Figures 2c, Figure 3), significantly reducing the transverse dimensions of the acicular α-phase. Its orientation becomes more ordered compared to the sample with the shortest exposure time (Figure 2a).

The research results showed that when the exposure time at the brazing temperature (Tp = 900 °C) is increased from 15 to 60 minutes, the width of the brazed seam increases by 2 to 2.5 times (Figure 2, a-c).

The mechanical tests of the brazed joints correlate with the metallographic findings and demonstrate that as the exposure time increases from 15 to 60 minutes, the homogeneity of the structure improves, and the shear strength increases by 15% (from 430 to 500 MPa).

4 CONCLUSIONS

The influence of holding time during vacuum high-temperature brazing on the structure and mechanical properties of TA15 alloy joints, using Ti-Zr-Me(Cu, Ni) filler metal, was investigated.

It was determined that preliminary mechanical treatment of the base metal increases the spread area of the filler metal by 10-20% compared to samples subjected to etching in a mixture of acids (HNO_3 + HF). Increasing the heating temperature from 900 °C to 950 °C also enhances the spread area by 10%.

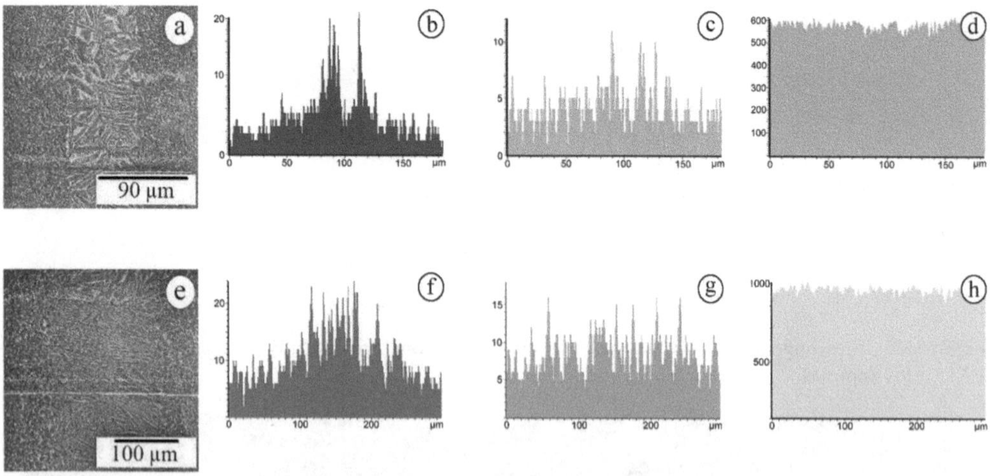

Figure 3. The microstructure of the brazed seam (a, e) and the distribution of alloying elements (b, f – Cu; c, g – Ni; d, h – Ti) in the brazed seam obtained at different holding times (min): 15 (a-d) and 60 (e-h).

Micro-X-ray spectral studies revealed that extending the exposure time from 15 to 60 minutes (at Tb = 900 °C) approximately doubles the width of the brazed joints (from 70 to 162 μm). Additionally, it improves the homogeneity of the joint structure and increases the shear strength by about 15% (from 430 MPa to 500 MPa).

REFERENCES

Gillespie, F. 1999. Titanium in fatigue critical military aircraft structures. *Fatigue Behavior of Titanium Alloys*, ed. by R.R. Boyer, 299–306. TMS.

Huang, S., Sun, B. & Guo, S. 2021. Microstructure and property evaluation of TA15 titanium alloy fabricated by selective laser melting after heat treatment. *Optics & Laser Technology* 144: 107422.

Huai, J., Guo, W. & Li, T. 2013. Brazing technology and joint properties of TA15 Titanium Alloy. *Journal of Aeronautical Materials* 33(3): 64–68.

Jing, Y., Su, D., Yue, X., Britton, T. B. & Jiang, J. 2017. The development of high strength brazing technique for Ti-6Al-4V using TiZrCuNi amorphous filler. *Materials Characterization* 131: 526–531.

Jing, Y., Xiong, H., Shang, Y. & Cheng, Y. 2020. Simulation on Ti-based filler and vacuum brazing for TA15 alloy. *Weld World* 64: 1261–1268.

Jing, Y., Xiong, H., Shang, Y., Wang, J., Cheng, Y. & Jiang, J. 2020. Design TiZrCuNi filler materials for vacuum brazing TA15 alloy. *Journal of Manufacturing Processes* 53: 328–335.

Jing, Y., Yue, X., Gao, X., Su, D. & Hou, J. 2016. The influence of Zr content on the performance of TiZrCuNi brazing filler. *Materials Science and Engineering: A*, 678: 190–196.

Muhrat, A., Puga, H., Barbosa J. & Puga, H. 2018. Low-Temperature Brazing of Titanium Using Al-Based Filler Alloys. *Advances in Materials Science and Engineering* 2018(1): 1–16.

Pang, S., Sun, L., Xiong, H., Chen, C., Liu, Y., Li, H. & Zhang, T. 2016. A multicomponent TiZr-based amorphous brazing filler metal for high-strength joining of titanium alloy. *Scripta Materialia* 117: 55–59.

Rabinkin, A., Liebermann, H., Pounds S., Taylor, J., Reidinger, F. & Siu-Ching, L. 1991. Amorphous Ti-Zr-base Metglas brazing filler metals. *Scripta Met. Et Mater* 25: 399–404.

Shapiro, A.E. 2017. Brazing of Conventional Titanium Alloys. *ASM Handbook vol. 6: Welding, Brazing, and Soldering*, ASM International.

Shapiro, A.E. & Flom, Y.A. 2007. Brazing of titanium at temperatures below 800°C: review and prospective applications. Retrieved from http://www.titanium-brazing.com/publications/DVS-Manuscript_1020-Copy2-19-07.pdf (Accessed 10 Sep. 2024).

Wei, X., Zhu, X., Wang, Z., Ran, X., Li, Z., Chen, R., Du, Z., Zhang, X., Singh, H., Zhang, J. & Lu, X. 2024. Comparative study of Ti-6.5Al-2Zr-1Mo-1V (TA15) alloy fabricated by selective laser melting and laser metal deposition. *Optics & Laser Technology* 176: 111052.

Welding and Related Technologies – Krivtsun et al. (Eds)

Evaluation of the efficiency of energy transfer of a laser-plasma heat source to metal during welding of carbon and stainless steel

V.M. Korzhyk, S. Gao & X. Wang
China-Ukraine Institute of Welding, Guangdong Academy of Sciences, Guangdong Provincial Key Laboratory of Material Joining and Advanced Manufacturing, Guangzhou, China

V.Yu. Khaskin, I.M. Klochkov, Ye.V. Illiashenko, O.I. Bushma & O.Ye. Bozhok
E.O. Paton Electric Welding Institute of the NAS of Ukraine, Kyiv, Ukraine

O.V. Dolyanivska
National Technical University of Ukraine «Igor Sikorsky Kyiv Polytechnic Institute», Kyiv, Ukraine

ABSTRACT: The effectiveness of energy absorption during laser, plasma, and laser-plasma welding of carbon and stainless steels using fiber laser radiation was investigated. It is shown that during laser-plasma welding, the energy absorbed by the metal is 5-6% higher than the arithmetic sum of absorbed energies under the separate action of laser radiation and plasma arc. This is mainly due to an increase in arc voltage in the process of laser-plasma welding compared to plasma welding.

1 INTRODUCTION

In welding steels and alloys with mutual action of focused laser radiation and arc or plasma-arc heat source into a common weld pool an increase of the cross-sectional area of the weld (approximately up to 4 times) and of penetration depth is observed. This effect is called synergetic and it may arise due to additional compression of the arc heat source to the dimensions of the plume of metal vapours ionized by laser radiation (Krivtsun et al. 2016, Möller & Thomy 2013). In addition to the action of the laser heat source on the arc source, the influence of the arc source on the laser one is in place. It consists in an increase of the part of laser radiation absorbed by the weld pool metal, earlier melted by the arc source. It is currently believed that at application of solid-state lasers (Nd-YAG, fiber and disc) the mechanism of compression of the arc energy source due to the action of the laser source makes the most significant contribution to creation of a synergetic effect in the processes of laser-arc or laser-plasma welding (Reisgen et al. 2016). Much less attention is given to increase of the part of absorbed radiation. The synergetic effect is predominantly evaluated by the geometry of the produced welds (Unocic & DuPont 2004). Here, the authors' opinions differ as to the part of energy, which is additionally generated at interaction of both the energy sources in the weld pool.

2 OBJECTIVE

The objective of the work is determination of the energy manifestation of the action of synergetic effect, caused by mutual action of the radiation of fiber laser and compressed plasma arc on the carbon and stainless steels during welding.

DOI: 10.1201/9781003518518-12

The following investigations were performed to achieve the defined objective. Laser, plasma and laser-plasma processes were used to deposit beads in samples of Q235 carbon steel and AISI304 stainless steel 6 mm thick with the following calorimetry to determine the part of thermal energy applied to the sample. Derived data were compared with laser and plasma process power that allowed evaluation of the effectiveness of energy input into the metal being welded in the considered welding processes.

3 RESEARCH AND DISCUSSION

Effectiveness of absorption of the laser energy (or energy of another welding source) by the metal can be characterized through effectiveness of energy transfer η_{en}, which is given by the formula proposed in work (Unocic & DuPont 2004):

$$\eta_{en} = Q/Pt \tag{1}$$

where Q is the energy absorbed by the metal, J; P is the initial power of radiation of the laser (or arc welding source), W; t is the duration of action of laser radiation (or welding source in general), s.

Calorimetry method was used to determine the quantity of energy Q absorbed by the metal (Gu et al. 2022, Gu et al. 2022, Skorokhod et al. 1995). For this purpose, initial T_1 and final T_2 (after placing the treated sample into the calorimeter) temperature of water in the calorimeter was determined. Absorbed energy Q was calculated by the following formula:

$$Q = (c_{H2O} \cdot m_{H2O} + c_{MK} \cdot m_{MK}) \times (T_2 - T_1) \tag{2}$$

where c_{H2O} = water heat capacity, kJ/kg·K; m_{H2O} = water weight, kg; c_{MK} = heat capacity of the material, from which the calorimeter is made, kJ/kg·K; m_{MK} = calorimeter weight, kg; T_1 = initial water temperature, K; T_2 = final water temperature, K.

Absorbed power, kW:

$$P_{ads} = Q/t \tag{3}$$

where t = time of action of the heat source on the sample, s (of the order of 11.5 s). The effectiveness of energy transfer can be represented in the following form:

$$\eta_{en} = P_{ads}/P \tag{4}$$

where P = initial power, kW.

The determined values of absorbed energies, mode parameters and calculated values of energy effectiveness of the welding processes are given in Table 1. Investigations of the welds were conducted using the procedure of (Korzhyk et al. 2021, Bell & Sonon 1976, Kvasnytskyi et al. 2020, Zamanzade et al. 2016, Malinina 2002).

One can see from comparative analysis of the obtained data that the actual energy absorbed by the metal in the hybrid welding process exceeds the algebraic sum of absorbed energies of the laser and plasma by 5-6 %. This additional part of energy absorbed by the metal being welded formed due to manifestation of the synergetic effect in laser-plasma welding owing to a combined action of the laser and plasma energy sources into a common weld pool. The effect of additional energy generation can be explained by formation of a keyhole in the weld pool under the impact of both the energy sources that promotes elongation of the plasma arc immersed into this keyhole. Increase of arc voltage is exactly what promotes energy increment.

Despite generation of additional energy, the effectiveness of energy input into the weld pool at laser-plasma welding decreases compared to laser welding. This is attributable to the fact

Table 1. Results of determination of the effectiveness of energy transfer at laser, plasma and laser-plasma welding of Q235 and AISI304 steels at the speed of 0.6 m/min.

Material	Welding process	P_{las}, kW	I, A	U V	T_1, °C	T_2, °C	Q, kJ	P_{ads}, kW	η_{en}, %
	Laser	0.8	-	-	22.9	25.8	7.20	0.63	78.3
	Plasma	-	80	24.1	20	25	12.42	1.08	56.0
Q235	Laser-plasma	0.8	80	26	22.8	31.2	20.86	1.81	63.0
	Total energy Q, kJ	-	-	-	-	-	19.62	-	-
	Laser	0.8	-	-	22	25	7.45	0.65	80.9
	Plasma	-	80	24.5	22	27	12.42	1.08	55.1
AISI304	Laser-plasma	0.8	80	26.3	22.4	30	21.11	1.84	63.2
	Total energy Q, kJ	-	-	-	-	-	19.87	-	-

that the plasma component of the laser-plasma process has the main role in increase of the melt volume by 2 – 4 times, and this component has a lower efficiency of energy input. One can see from Table 1 that the arc voltage at laser-plasma welding becomes higher compared to plasma welding. As is noted in work (Krivtsun et al. 2016), increase of penetration depth occurs through increase of current density, which is due to interaction of focused radiation with arc plasma, that, in its turn, causes an increase of the electromagnetic force, which transfers the molten metal from the surface to the bottom of the weld pool, thus enhancing the effectiveness of convective heat transfer into the pool depth.

4 CONCLUSIONS

It was found that the energy absorbed by carbon and stainless steels during their laser-plasma welding using fiber laser radiation, is approximately the same and is by 5-6% higher than the arithmetic sum of energies absorbed at independent action of laser radiation and plasma arc. This additional energy generation is due to the manifestation of the synergetic effect at interaction of the laser and plasma components of the process in the weld pool, and it is mainly due to increase of arc voltage during laser-plasma welding compared to plasma welding.

ACKNOWLEDGMENTS

The work was funded within the following programs:

1. The National Key Research and Development Program of China [grant number 2020YFE0205300], China.
2. GDAS'Project of Science and Technology Development (2020GDASYL-20200301001), China.
3. Research projects of China [G2023030004L], China.

REFERENCES

Krivtsun I. et al. 2016. Modelling of weld pool phenomena in tungsten inert gas, CO_2-laser and hybrid (TIG+CO_2-laser) welding. *Journal of Laser Applications*. 28(2): 022406.

Han, J. et al. 2022. Minimizing defects and controlling the morphology of laser welded aluminium alloys using power modulation-based laser beam oscillation. *Journal of Manufacturing Processes*. (83): 49–59.

Möller F. & Thomy C. 2013. Interaction effects between laser beam and plasma arc in hybrid welding of aluminum. *Physics Procedia*. (41): 81–89.

Reisgen, U. et al. 2016. Experimental research of hybrid welding processes in combination of gas tungsten arc with CO_2- or Yb:YAG laser beam. *Journal of Laser Applications*. (28): 022402.

Unocic, R. R. & DuPont, J. N. 2004. Process efficiency measurements in the laser engineered net shaping process. *Metallurgical and Materials Transactions B*. 35(1): 143–152.

Gu, Y. et al. 2022. Corrosion resistance of 316 stainless steel in a simulated pressurized water reactor improved by laser cladding with chromium. *Surface and Coatings Technology*. (441): 12853.

Gu, Y. et al. 2022.Stress-assisted corrosion behaviour of Hastelloy N in FLiNaK molten salt environment. *Materials Degradation*. 6(1): 90.

Skorokhod, A.Z. et al. 1995. The effect of mechanical pretreatment of polyethylene terephthalate powder on the structural and mechanical properties of coatings made from it. *Mechanics of Composite Materials*. 30(4): 328–334.

Korzhyk, V. et al. 2021. Comparing features in metallurgical interaction when applying different techniques of arc and plasma surfacing of steel wire on titanium. *Eastern-European Journal of Enterprise Technologies*. 4(12): 6–17.

Bell F.C. & Sonon D.E. 1976. Improved metallographic etching techniques for stainless steel and for stainless steel to carbon steel weldments. *Metallography*. 9(2): 91–107.

Kvasnytskyi, V. et al. 2020. Designing brazing filler metal for heat-resistant alloys based on ni al intermetallide. *Eastern-European Journal of Enterprise Technologies*. 6(12): 6–19.

Zamanzade M. et al. 2016. Review on the properties of iron aluminide intermetallics. *Crystals*. (6): 1–29.

Malinina R.I. (ed.). 2002. *Practical metallography*. Moscow: Internet engineering.

Welding and Related Technologies – Krivtsun et al. (Eds)

The nanomodification influence on the microstructure formation of low alloy steel welds metal

V.V. Holovko, V.A. Kostin, V.V. Zhukov & S.M. Stepanyuk
Paton Electric Welding Institute, Kyiv, Ukraine

ABSTRACT: Studies on the nano-sized refractory particles inoculation of oxides Al_2O_3, TiO_2, MgO, ZrO_2 into the welding bath were performed. The oxides inoculation with a low level of mismatch to the δ-Fe lattice and increased wetting with liquid iron (MgO, ZrO_2) into the weld pool contributes to the dendrite's growth. Metal modification with oxides Al_2O3, TiO_2 contributes to an increase in the polygonal ferrite hardness and an increase in the upper bainite content in the weld's microstructure, growth the low-alloy steel welds strength indicators.

1 INTRODUCTION

Despite the rapid development and wide use of various composite and polymer materials in recent times, steel still remains the main material in the manufacture of welded metal structures in various the economy branches. The most noticeable growth rates are characteristic for high-strength low-alloy (HSLA) steels. The expansion of objects in which HSLA steel is used is accompanied by the constant improvement of their production technology (Varshney & Sangal 2016). Due to the fact that the manufacture, repair and restoration of HSLA steels structures made is associated with the wide use of welding processes, the problem of increasing the welds metal reliability in accordance with the metal mechanical properties level is highly relevant.

The expansion of the HSLA steels use is associated with an effective strength, plasticity and viscosity indicators combination, which are achieved as a result of complex manufacturing processes (Halfa 2014). Welding processes have a significantly smaller technological approaches list to solving the problem of ensuring a complex of mechanical properties of welded joints. The entire steel welding technological process is based on understanding the mechanisms of weld metal structure controlling and the structure parameters (size of grains, phases, inclusions) in the structure of the weld metal. The more control is given over this process, the higher the level of their mechanical properties can be obtained. One of the common methods of increasing the weld metal toughness without reducing their strength is the weld metal modifying process (Kostin & Zhukov 2020).

The weld metal modification process has long been used both in the production of steel castings and in welding in order to influence the metal mechanical properties. Recently, publications have appeared in the scientific and technical literature, which show the inoculating nano-sized particles expediency of refractory compounds into the metal melt to modify the metal in the crystallization and microstructure formation process (Hsiung & Fluss 2010). Works have been carried out in which the nano-sized inoculants influence on the welds metal properties has been shown (Holovko & Stepanyuk 2012).

2 MATERIALS AND TECHNIQUE

The research was carried out on HSLA steel welds metal (a yield strength up to 650 MPa), which were performed according to the arc welding method with a shielding gas (82% Ar, 18% CO_2) and use the "metal core" type flux-cored wire with a 1.6 mm diameter at a 200 A direct current, a 30 (±2) V arc voltage with a 21 (±2) kJ/cm input energy. The energy level at the non-

DOI: 10.1201/9781003518518-13

metallic inclusion boundary with a metallic phase can be estimated through the wetting angle. Therefore, refractory compounds with different wetting levels by the metal phase were chosen as inoculants (Table 1). The mechanical properties determination results are shown in Table 2.

Table 1. Chemical composition of the experimental welds metal.

	Chemical composition. wt. %										
Weld metal	C	Si	Mn	S	P	Cr	Ni	Mo	Al	Ti	Inoculant
HH-0	0,042	0,34	1,19	0,021	0,020	0,10	2,13	0,28	0,028	0,029	Base
HH-22	0,035	0,40	1,24	0,016	0,021	0,11	1,97	0,27	0,031	0,017	TiO_2
HH-23	0,034	0,42	1,40	0,017	0,023	0,15	2,15	0,29	0,032	0,015	Al_2O_3
HH-24	0,031	0,22	1,11	0,025	0,024	0,14	1,85	0,29	0,023	0,030	MgO
HH-25	0,033	0,22	1,05	0,024	0,024	0,12	2,02	0,30	0,024	0,031	ZrO_2

Table 2. Mechanical properties of welds metal.

	Rm	Re	A	Z	KCV, J/cm^2 at T, °C				
Weld metal	MPa		%		+ 20	0	– 20	– 40	Inoculant
HH-0	693,0	605,0	14,5	48,4	97,0	87,0	75,0	53,0	Base
HH-22	708,7	636,4	19,3	56,7	84,6	71,7	60,0	50,0	TiO_2
HH-23	728,2	621,4	17,5	54,4	82,1	58,3	50,4	35,8	Al_2O_3
HH-24	644,5	586,0	18,6	59,9	102,9-	-	69,2	60,0	MgO
HH-25	621,6	532,2	19,5	65,0	119,6	-	72,9	64,6	ZrO_2

To determine the nature of the non-metallic inclusions, influence on the welds metal microstructure formation, a powder wire 1.6 mm diameter was introduced into the "cold" part of the welding bath. The wire core contained a mixture of 10% refractory compounds particles with a size of 0.040...0.200 mm and 90% of iron powder according to DSTU 9849. The obtained results were compared with the data of weld metal samples obtained when welding with flux-cored wire that did not contain inoculants (weld metal HH-0).

The nature of structural transformations in the alloyed welds metal was studied by the simulating method the welding thermal deformation cycle using the Gleeble 3800 complex equipped with a high-speed dilatometer. The cooling curves corresponded to the Newton-Richmann dependence and corresponded to the metal cooling rates 5...45 °C/s in the 500...600 °C.

3 RESEARCH RESULTS

The microstructure of the investigated welds metal, which were modified with various refractory metals oxides (TiO_2, Al_2O_3, MgO, ZrO_2), was investigated by the optical metallography methods, Figure 1. Quantitative metallography methods were used to determine the dendrites and primary structure grains sizes and the structural components content in the welds (Table 3).

The primary austenite structure is the basis of the welds metal secondary structure formation processes, which corresponds to the entire complex of its mechanical properties. The austenite transformation occurs in the high temperature range Ac_3–Ac_1, which depends on the weld metal cooling rate.

The dilatometric studies data, as well as the volume fraction transformation data dependence on the temperature, indicate a monotonous the phase transformation course, when it is not possible to distinguish the intermediate phases occurrence points, except for the beginning and end the transformation points, for all cooling rates on austenite decomposition diagram.

In addition to the primary austenite grains morphology, the secondary structure formation of the low-alloy steels weld metal is influenced by the bainite beginning and end transformation temperature. The bainite formation kinetics was evaluated by the B_{50} indicator, which corresponds to the 50% the bainite phase formation temperature. In order to identify the

Table 3. Share (%) of welds metal microstructural components (*).

Welds metal	Ferrite					Bainite		Other
	AF	PF	IGF	GF	WF	Upper	Lower	
HH-0	8	5	8	2	15	40	17	5
HH-22	5	5	0	0	32	43	10	5
HH-23	2	2	0	0	30	48	11	7
HH-24	32	10	0	10	7	15	21	5
HH-25	30	15	2	6	7	10	25	5

* Ferrite: AF - acicular; PF - polygonal; IGF - intragranular; GF - globular; WF – Widmannstätten.

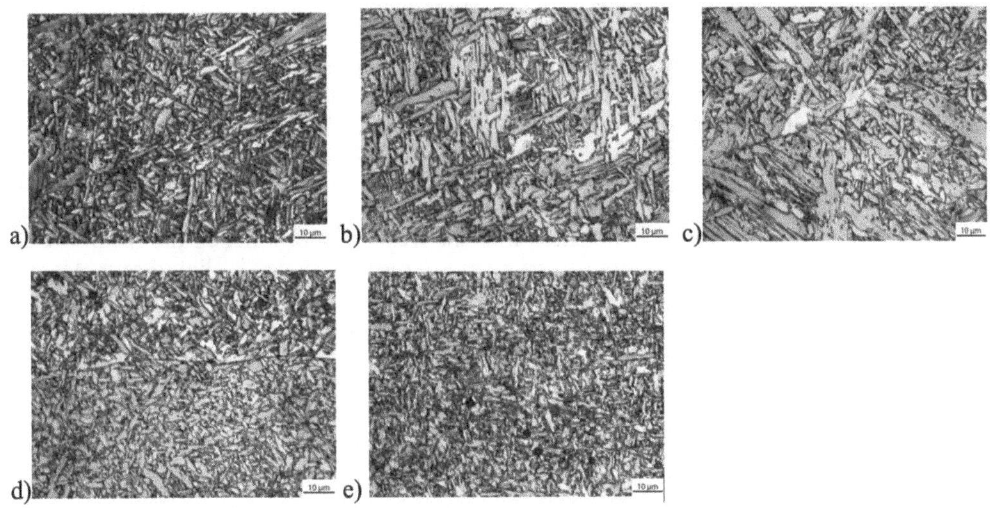

Figure 1. Microstructure of welds metal samples: a- without inoculants; and modified with inoculants: b - TiO_2; c - Al_2O_3; d - MgO; e - ZrO_2.

factors that influence these processes, a comparative analysis of the dilatograms phase transformation was performed, the results are shown in Table 4.

Table 4. Parameters (*) of austenite decomposition diagrams in the weld metal structure.

Weld metal	Transformation temperature, °C						
	Ac3	Ac1	ΔA	Bs	B_f	ΔB	B_{50}
HH-0	876	627	249	677	421	256	577
HH-22	875	660	215	653	417	236	568
HH-23	871	668	203	656	434	222	567
HH-24	870	628	242	689	452	237	586
HH-25	853	604	249	684	456	228	577

* Ac3 - the beginning of the austenite formation; Ac1 - the beginning of the austenite breakdown; ΔA is the range of austenite decay; Bs is the beginning of the bainite transformation; Bf – end of bainitic transformation; ΔB – range of bainite transformation; B50 is the formation temperature of 50% bainite

4 DISCUSSION

The results of dilatometric studies showed that the aluminum and titanium oxides inoculation in the welding pool reduce the austenite decomposition temperature range by increasing the temperature of its completion (Table 4). Modification with magnesium and zirconium oxides contributes to the formation of more massive dendrites during metal crystallization. When magnesium and zirconium oxides nanoparticles are introduced into the welding pool, the bainitic transformation beginning and end temperature increases significantly (Table 4). An increase in the dendrites size and a decrease in the B_f temperature affects the carbon diffusion reduction from the austenite grains in the $\gamma \rightarrow \alpha$ transformation process, which contributes to increasing the cementite formation in the body and a decrease in their allocation at the ferrite grains boundaries. The such metallurgical reactions development probability is confirmed by the increased lower bainite content in the weld metal structure (Table 3) and the reducing the polygonal ferrite hardness level in comparison with the welds metal modified with Al_2O_3 and TiO_2 oxides.

The weld metal modification with aluminum and titanium oxides is accompanied by a decrease in the temperature at which the bainite transformation ends. As a result, the carbon amount that diffuses to the grain boundaries increases, the polygonal ferrite hardness level increases and the upper bainite content increases.

Changes in the of the welds metal microstructure composition correspond to the results of determining their mechanical properties. An increase in the polygonal ferrite hardness and the upper bainite content in the microstructure composition when modifying the weld metal with aluminum and titanium oxides is accompanied by an increase in strength properties (Table 2). The welds metal modification with magnesium and zirconium oxides helps to increase the lower bainite content in the microstructure composition and increase the level of metal viscosity indicators (Table 2).

It can be seen that the oxides inoculated into the welding pool differ in the inconsistency level of the crystal lattice parameters with iron and the wetting index by liquid iron. Magnesium and zirconium oxides, which are well wetted by molten iron, and whose crystal lattice size is close to the δ-Fe parameters, should be iron crystals nucleation centers at higher temperatures compared to the welds metal inoculated with aluminum and titanium oxides. This explains the difference in the dendrites size, which was established during the welds metal metallographic analysis. Reducing the inoculated into the weld pool refractory oxides size to the size of the growing dendrites top in the steel melt increases the modification efficiency.

5 CONCLUSIONS

It was established. The oxides inoculation into the weld pool with a low level of mismatch to the δ-Fe lattice and increased wetting with liquid iron (MgO, ZrO2) contributes to the dendrites growth that are formed during melt crystallization. The metal modification with MgO, ZrO_2 oxides reduces the polygonal ferrite hardness and increase the lower bainite content in the welds metal microstructure. The metal modification with Al_2O3, TiO_2 oxides increases the polygonal ferrite hardness and increases the upper bainite content in the welds metal microstructure. As a result, the low-alloy steel welds metal strength and resistance to brittle fracture increases.

REFERENCES

Halfa, H. 2014. Recent Trends in Producing Ultrafine Grained Steels. *Journal of Minerals and Materials Characterization and Engineering* 2: 428–469.

Holovko, V.V. & Stepanyuk, S.N. 2012. Study of the effect of nanosized titanium carbides on the formation of the microstructure and properties of welds. *Physical chemistry of metals and metallurgy* 6: 68–75.

Hsiung, L.L. & Fluss, M.J. 2010. Formation mechanism and the role of nanoparticles in Fe-Cr ODS steels developed for radiation tolerance. *Physical Review* 82: 184103-1 - 184103-13.

Kostin V. & Zhukov V. 2020. Increase of Mechanical Properties of Weld Metal of High-Strength Low-Alloy Steels. *Springer Proceedings in Physics* 240:307–315.

Panasiuk, A.D. & Fomenko V.S. 1986. *Stability of non-metallic materials in melts*. Kyiv:Naukova dumka.

Varshney, A. & Sangal, S. 2016. Super strong and highly ductile low alloy multiphase steels consisting of bainite, ferrite and retained austenite. *Materials and Design* 95:75–88.

Welding and Related Technologies – Krivtsun et al. (Eds)

Laser welding of aluminum-copper joints using dynamic beam shape

A. Kovalevsky, D. Baulin & K. Zohar-Hauber
Israel Institute of Materials Manufacturing Technologies, Technion R&D Foundation, Technion City, Haifa, Israel

ABSTRACT: The paper presents a study on dissimilar laser welding of pure copper and Aluminum 3003 alloy. During this research conduction mode and keyhole mode were investigated. The microstructure of welded seams and the heat-affected zone were analyzed. Results demonstrated no mixing of metals and a significant presence of intercrystallite composites in the conduction mode. The use of a dynamic beam shape, characterized by two points with varying energy densities, resulted in uniform heat distribution at the copper-aluminum interface and high-strength joint. The tensile strength of the welded pipes ranged from 107-112 MPa, representing 75-80% of the base metal's strength. Additionally, the joints successfully passed a leak test at 43 atmospheres.

1 INTRODUCTION

Dissimilar metal (Al+Cu) joints can significantly increase the flexibility in engineering design and production processes. Also, the electrical, mechanical, thermal, and corrosion resistance properties of such materials may improve from different combinations of metals (Von, Blatter 1995). While copper possesses an excellent electric conductivity, its high density restricts the possibilities for lightweight designs. Furthermore, the comparatively high market price of copper leads to a cost increase for electronic constructions. This is the reason why it is intended to join copper with aluminum, which features similar electric properties at a reduced market price and a lower specific density for various industries, such as electronics, automotive, power systems.

2 CHALLENGES DURING AL – CU WELDING

Dissimilar metal joining of copper and aluminum is a well-known problem and has been the subject of investigations for decades. The main problem while joining this dissimilar combination is the formation of brittle intermetallic phases that degrade the electrical and mechanical properties of the joint. Very high intensity is required to join copper and aluminum sheets with a laser. The intermixture of copper and aluminum is vital for joint ductility. Besides, the laser welding of copper and aluminum poses multiple challenges including reflectivity, high thermal conductivity, and large differences in thermal expansion and melting points (Zhou et al. 2017). Another main challenge is the low absorption of aluminum and especially copper at the laser wavelength of 1 μm (Stritt et al. 2014). For laser welding, the spatial beam oscillation is effective in increasing the weld seam width and decreasing the depth (Fetzer et al. 2016). Controlling the depth of fusion in the Al-Cu system close to the interface is beneficial as mixing can be kept to a minimum (Shmallen et al. 2016). Welding with laser beam oscillation (or laser wobbling) when the laser beam is put in oscillation regarding to the molten puddle. Such oscillation, indicated as secondary motion, can occur with linear or curvilinear (circular, elliptic, or other) motion. This motion superimpose to the feeding motion of the laser beam along the welding path. Such oscillation technique becomes very promising especially when we refer to the use of fiber lasers, characterized by a higher quality of the laser beam that makes it more focusable on small sizes. Wobbling of the laser beam enlarges

DOI: 10.1201/9781003518518-14

the beam/material interaction area and the weld width while decreasing the heat input required during welding (Wang et al. 2019). By wobbling the beam in highly reflective materials, such as copper, the local temperature of the material increases and so does its absorptivity. The dynamic movement of the laser can have different shapes and its controlled oscillation enables better temperature management during the process, which results in less steep heat, cooling rates, and thermal gradients. Thus, laser oscillation can improve the process by minimizing welding defects, without deleterious effects on the microstructure for Cu/Al dissimilar joints (Kraetzsch et al. 2011).

3 MATERIALS AND METHODS

The experiments were conducted using two pipes from pure copper and Al3003 alloy with thicknesses of 1 mm and diameters of 16 and 16.5 mm respectively. A copper tube was inserted into an aluminum tube and a laser beam welded the samples together, making an overlap joint (Figure 1). Prior to welding, the samples were cleaned by grinding with a wire brush and subsequently washed with acetone and ethanol to remove oxide films and grease. Argon with 99.99% of purity was used as a shielding gas to protect the welding zone, with a flow rate of 15 L/min. Obtained cross-sectional profiles during the welding were observed by an optical microscope. Tensile testing was performed according to the ISO 6892-1:2016.

In the experiments we used a high-power 14Kw fiber CW laser (Civan) with 1064 nm wavelength and dynamic beam shape feature. This technology allows dynamically tailor beam parameters to the required application. With the technology of CIVAN Optical Phase Array (OPA), multiple laser channels are electrooptically controlled to combine or interact with each other. For this study, a dynamic beam shape consisting of two points with different energies (Figure 1) was used. The laser machine and welding scheme is shown in Figure 1.

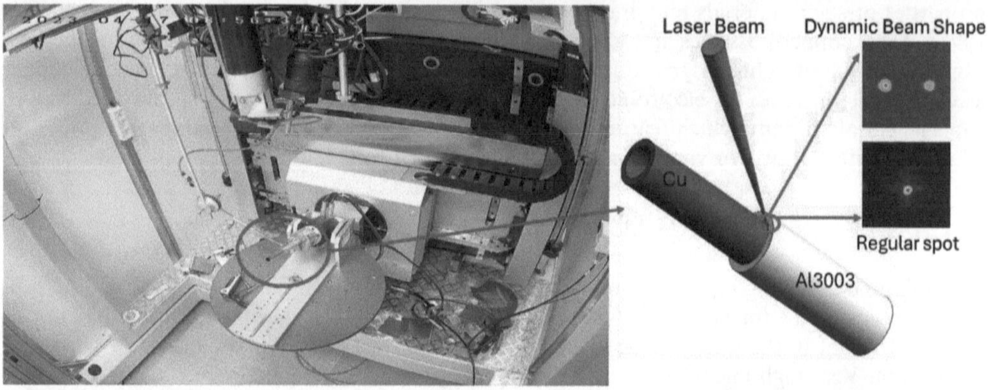

Figure 1. Welding cell, scheme of welding process, regular and dynamic beam shape (2 dots).

4 RESULTS AND DISCUSSION

During the study, two different welding modes (conduction and keyhole) were investigated. Dissimilar circular welding pipes were made of Al3003 alloy and pure copper. Welding parameters for both are presented in Table 1.

Keyhole mode in laser welding refers to a process where the laser creates a small, deep vapor cavity (keyhole) in the material being welded. This happens when the laser power density is high enough to vaporize the material, forming a narrow, deep hole surrounded by molten metal. The keyhole is maintained by the pressure of vaporized material and is stabilized by surface tension and the flow of molten material. Optical and SEM images of weld are shown in Figure 2.

Image analysis shows significant mixing of metals and their mutual penetration into each other. This led to the formation of different Al-Cu phases as AlCu, Al_2Cu, Al_4Cu_9 and to

Table 1. Welding parameters.

Mode	Keyhole	Conduction
Power, W	9000	5000
Speed, Rpm	30	15
Focus, mm	0	-30
Beam shape	Point	2-points
Frequency, KHz	200	200
Shielded Gas	Argon	Argon

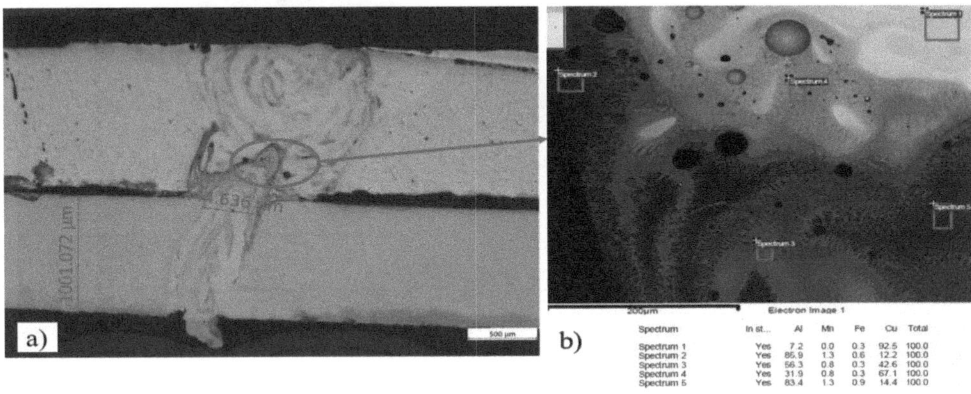

Figure 2. Microscopical investigation of welded joint (keyhole mode): a – optical microscope image, b – electron microscope image and phase analysis.

a significant amount of intermetallic composites. Also, considerable mixing of metals led to defects in the weld - pores and cracks. Using this mode and various welding parameters, none of the experiments shown satisfactory results.

Conduction mode in laser welding occurs when the laser power density is relatively low, and the energy from the laser beam is absorbed by the surface of the material rather than penetrating deeply. In this mode, the heat is conducted through the material primarily by thermal diffusion, causing surface melting without creating a deep keyhole, as in keyhole mode.

During the experiment, a dynamic beam in the form of two dots was used to distribute the spot energy on the metal surface evenly. The dot with more concentrated energy was directed at copper since it has a higher melting point. The use of dynamic beam shape and conductive mode resulted in much better results than with keyhole mode. Image analysis shows low mixing, narrow intermetallic zone (less than 150 µm), and absence of welding defects (Figure 3).

Tensile testing was performed at a 5 mm/min constant loading rate using an Instron 3369 universal testing machine (equipped with 50 kN Instron loadcell).

As depicted in Table 2, the average tensile strength of the welds achieved through conventional welding and employing a dynamic beam shape in the form of two points was 105-110 MPa. These values corresponded to approximately 75 % of the base metal strength (Al3003 - 147 MPa). Welding was carried out without any filler or welding wire. The leak test showed that all welded joints passed the test at a pressure of 43 atmospheres.

5 CONCLUSIONS

In this paper were investigated conduction and keyhole modes of tube laser welding of dissimilar metals - aluminum alloy 3003 and pure copper. The conduction mode weld was performed at a power of 9 kilowatts and the keyhole mode was performed at a power of 5 kilowatts. It is shown that the mixing of metals in the conduction mode is significantly less than in the keyhole mode. In the conduction mode, the area and size of intermetallic composites are smaller, as well

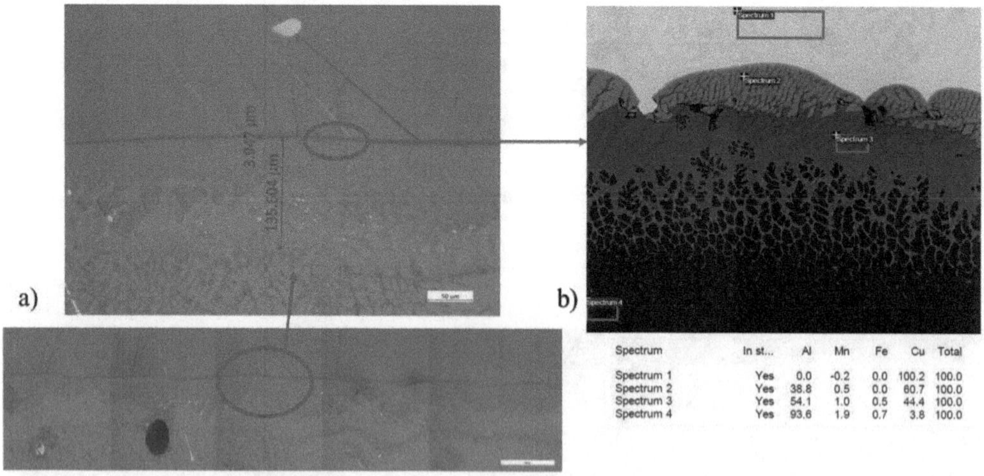

Figure 3. Microscopical investigation of the welded joint (conduction mode): a – optical microscope image, b – electron microscope image, and phase analysis.

Table 2. Tensile results.

Sample	Tensile stress at maximum load, Mpa
1	106
2	107
3	110
Al-Base	147
Cu-Base	246

as the number of cracks and pores. To ensure uniform melting of aluminum and copper during the welding, a dynamic beam shape in the form of two points with different energy concentrations was used. A more concentrated point was directed at copper, which is associated with higher reflectivity and a higher melting point. The welded joints showed an average tensile strength of 110 MPa, which is 80% of the base metal, and successfully passed a leak test at 43 atmospheres.

ACKNOWLEDGMENTS

The research was performed in the framework of the Laser for Advanced Material Processing Consortium under the support of the government of the State of Israel. For access to original data and additional information please contact the corresponding author.

REFERENCES

Fetzer, F. et al. 2016. Fine-tuned remote laser welding of aluminum to copper with local beam oscillation. *Phys. Procedia.* 83: 455–462.

Kraetzsch, M. et al. 2011. Laser beam welding with high-frequency beam oscillation: welding of dissimilar materials with brilliant fiber lasers. *Phys. Procedia.* 12: 42–149.

Schmalen P. et. Al. 2016. Evaluation of laser braze-welded dissimilar Al-Cu joints. *Phys. Procedia.* 83: 506–514.

Stritt, P. et al. 2014. Laser spot welding of copper-aluminum joints using a pulsed dual wavelength laser at 532 and 1064 nm. *Phys. Procedia.* 56: 759–767.

Von, A. & Blatter. A. 1995. *Laser-beam interactions with materials: physical principles and applications.* Berlin: Springer.

Wang, Z. et al. 2019. Laser beam oscillating welding of 5A06 aluminum alloys: microstructure, porosity and mechanical properties. *Opt. Laser Technol.* 111: 58–65.

Zhou, X. et al. 2017. Microstructures and mechanical behavior of aluminum-copper lap joints. *Mater. Sci. Eng.* 705(8): 105–113.

Welding and Related Technologies – Krivtsun et al. (Eds)

Using an automatic monitoring system to automate the management of welding and 3D printing processes

V.M. Korzhyk, S. Gao & X. Wang
China-Ukraine Institute of Welding, Guangdong Academy of Sciences, Guangdong Provincial Key Laboratory of Material Joining and Advanced Manufacturing, Guangzhou, China

O.M. Voitenko, V.Yu. Khaskin, I.M. Klochkov & Ye.V. Illiashenko
E.O. Paton Electric Welding Institute of the NAS of Ukraine, Kyiv, Ukraine

I.O. Skachkov
National Technical University of Ukraine «Igor Sikorsky Kyiv Polytechnic Institute», Kyiv, Ukraine

ABSTRACT: The concept of creating a multi-information system for collecting and analyzing information for automatic monitoring of welding and 3D printing processes is proposed, which is based on the application of neural networks with self-learning and information databases with relevant parameters of welding processes. Based on the action of such neural networks, feedback for automated control of these processes will be implemented in the system, which will allow creating an automatic monitoring system capable of controlling welding and 3D printing processes in real time.

1 INTRODUCTION

Modern welding processes should have a high level of automation due to increased complexity of the tasks facing them. For instance, laser-arc welding processes usually have a high speed and require rather complex equipment, operating exclusively in the automatic mode (Acherjee 2018, Gu et al. 2022, Han et al. 2022, Zhang et al. 2022, Gu et al. 2022). 3D printing processes are performed by layer-by-layer additive deposition along complex trajectories that also requires complete automation (Korzhyk et al. 2024). Systems of welding process automation include not only systems of movement of the tool (certain torches) by the assigned trajectory, but also real-time monitoring systems (Kah et al. 2014). Here, the quality and reliability of the welded joint, 3D structure, etc. largely depend on the quality and reliability of such systems.

The main problems of modern systems for automatic monitoring and control of welding processes are rational selection of the tracked process parameters, application of the required tracking sensors with high-enough response, as well as application of certain methods and algorithms for automated data processing. So, for arc welding processes the simplest and most common approach is controlling the arc voltage, which is directly proportional to arc length and stabilization of which reliably stabilizes process (Gu et al. 2011, Sydorets et al. 2017, Krivtsun et al. 2022). Such an approach however, does not allow controlling bead formation or minimizing porosity in real-time welding or deposition.

DOI: 10.1201/9781003518518-15

2 OBJECTIVE

The objective of the work is development of a concept of universal multi-information system for automatic monitoring of welding and 3D printing processes based on selection of the parameters of data collection, sensors for respective signals recording and application of neural networks for fast data processing and formation of process control feedbacks.

In order to achieve the defined objective, it is proposed to solve the following tasks. First, it is necessary to select the most informative channels of acquisition of diverse data as to running of the welding process. Such information usually belongs to electrical parameters, visual and acoustic monitoring. Then, it is necessary to select the methods of following and recording the required information, allowing for performance characteristics. After that, it is necessary to select the methods and algorithms for processing the collected information, formation of the respective control signal for each of the selected information channels and to realize the feedback in the form of the respective impact on the welding process. All this will allow creating a versatile multi-information system for automatic monitoring of welding and 3D printing processes.

3 RESEARCH AND DISCUSSION

Modern arc and plasma-arc welding processes were considered to achieve the defined purpose (Sahoo & Tripathy 2021, Kvasnytskiy et al. 2020, Korzhyk et al. 2021). Conducted analysis showed the following. For correct assessment of the situation as a whole, it is desirable to see both the zone of the welding source action, and the entire structure being welded. Recording of the shape of the welded structure or 3D object being built-up, will allow analyzing the deviations from the nominal shape and dimensions and making the required corrections. Laser scanners and CCD-videocameras will provide such information. Laser triangulation system will allow correlating the movement of the welding energy source with shaping of the structure as a whole and of its local zones. Temperature sensors (for instance laser pyrometers) will allow making a conclusion about the structure overheating, need for its local or general cooling (for instance, by compressed air), temperature fields in the zone of the welding source action, dimensions of the weld pool and hot zone behind it, etc. Monitoring the volt-ampere characteristics of welding equipment will provide information about the stability of its operation. Acoustic sensors will help evaluate the stability of the welding arc and of the welding or deposition process proper. To obtain information on the rate of gas flows (plasma forming, shielding, cooling, etc.) and the possibility to make corrections, it is necessary to use the respective flow meters. It is also possible to apply other sensors, in particular those related to individual features of a certain welding process and the metal being treated. It is also rational to apply welding arc control to implement the system for welding arc length stabilization. A system of spectroscopic diagnostics of welding arc plasma can be used to obtain more detailed information about the possibility of appearance of inner pores, defects of deposited metal bead formation, etc.

After recording certain parameters of the welding processes, it is necessary to interpret the signals from the respective sensors, i.e. compare the values of these signals with the information on admissibility or inadmissibility of maintaining a certain parameter within the current limits. For this purpose, it is necessary to create data bases with information on the signals from respective sensors, which will contain the data on the form of admissible or inadmissible signals. It is also necessary to create certain algorithms for processing the received signals and their comparison with those contained in the data bases. Such a task can be successfully performed by self-learning neural networks (Nguyen & Window 1990). Owing to their application, the self-learning process can be implemented when working with the information, and feedback can be realized on their base. Moreover, mathematical modeling (in particular, of the energy source nature and behaviour) can be used for correction of welding process parameters, and the results of the earlier performed mechanical testing can be taken into account.

It is proposed to perform recording of the parameters of welding or 3D printing processes using the multi-information system for data collection and analysis, which consists of a set of data measurement sensors, personal computer with a board for digital-analog inputs and configured data base for data storage and structuring (Figure 1). Such a system should be based on measurement of the parameters of welding or 3D printing process. PostgreSQL instrument can be used as the base for measurement result recording.

The proposed instrumentation for recording the considered signals (namely, signals of 3D scanning, laser triangulation measurement, measurement of temperature, welding arc voltage, video monitoring, spectroscopic diagnostics of welding arc plasma, gas flow rates, speed of movement of the working tool or treated structure, acoustic monitoring of the process, etc.) has rather fast response, usually from tens of nanoseconds to units of milliseconds. The proposed approach of application of self-learning neural networks will allow with a sufficiently fast response (up to units of milliseconds) processing the received signals and generating the respective feedback signals. All this will enable bringing the speed of the reaction to excitation in the weld pool or the welding source to 10 – 20 milliseconds. If we consider the speed of the processes, occurring in the weld pool, they are primarily related to the time of its existence. For plasma-arc processes of welding and 3D deposition, this time is equal to approximately 0.1 to 3.0 s. This is indicative of the suitability of the proposed approach of application of multi-information system for data collection and analysis, in order to create a system of automatic monitoring of the welding and 3D printing processes.

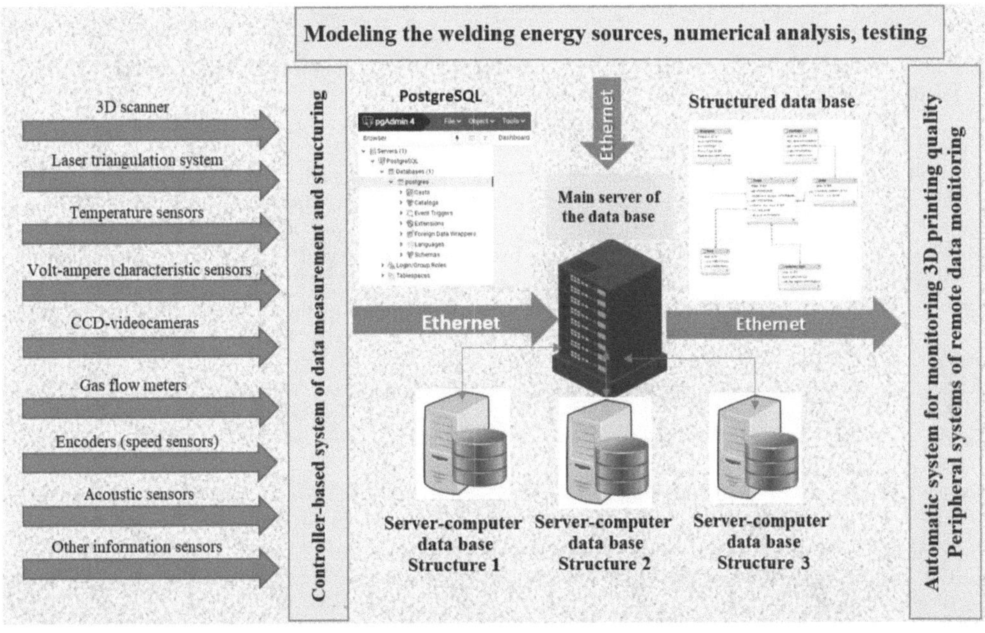

Figure 1. Hardware scheme of the system for automatic monitoring of the welding process quality, based on application of a multi-information system for data collection and analysis.

4 CONCLUSIONS

Thus, a multi-information system was proposed for data collection and analysis for automatic monitoring of the processes of welding and 3D printing, which is based on application of self-learning neural networks, contains data bases of the respective information on the welding processes and has feedback for automatic control of these processes. The proposed system includes the following subsystems: 3D scanning, laser triangulation measurement,

measurement of temperature, volt-ampere characteristics, monitoring/stabilization of welding arc voltage, videomonitoring, spectroscopic diagnostics of welding arc plasma, measurement of gas flow rates, measurement of the speed of movement of the working tool (treated structure), acoustic monitoring of the process, etc.

ACKNOWLEDGMENTS

The work was funded within the following programs:

1. The National Key Research and Development Program of China [grant number 2020YFE0205300], China.
2. GDAS'Project of Science and Technology Development (2020GDASYL-20200301001), China.
3. Research projects of China [G2023030004L], China.

REFERENCES

Acherjee, B. 2018. Hybrid laser arc welding: State-of-art review. *Optics & Laser Technology*. (99): 60–71.

Gu, Y. et al. 2022. Corrosion resistance of 316 stain-less steel in a simulated pressurized water reactor improved by laser cladding with chromium. *Surface and Coatings Technology*. (441): 12853.

Han, J. et al. 2022. Minimizing defects and con-trolling the morphology of laser welded aluminum alloys using power modulation-based laser beam oscillation. *Journal of Manufacturing Processes*. (83): 49–59.

Zhang, W. et al. 2022. Intergranular corrosion characteristics of high-efficiency wire laser additive manufactured Inconel 625 alloys. *Corrosion Science*. (205): 110422.

Gu, Y. et al. 2022. Stress-assisted corrosion behaviour of Hastelloy N in FLiNaK molten salt environment. *npj Materials Degradation*. 6(1): 90.

Korzhyk, V. et al. 2024. Features of the stress–strain State of 3D metal objects produced by additive microplasma deposition of the powder of a Fe–Cr–Ni–B–Si System. *Applied Sciences*. (14): 4159.

Kah, P. et al. 2014. Real-time weld process monitoring. *Advanced Materials Research*. (933): 117–124.

Gu, W. et al. 2011. An automatic control system based on arc voltage of TIG welding. *Advanced Materials Research*. (418-420): 1379–1382.

Sydorets, V. et al. 2017. Electrical characteristics of the equipment for the hybrid plasma-MIG welding. *58th Annual International Scientific Confererence on Power and Electrical Engineering of Riga Technical University, RTUCON 2017 - Proceedings, 2017-November*: 1–6.

Krivtsun, I. et al. 2022. Characteristics of high-frequency pulsed current arc with refractory cathode. *The Paton Welding Journal*. (3): 9–18.

Sahoo, A. & Tripathy, S. 2021. Development in plasma arc welding process: A review. *Materials Today: Proceedings*. 41(2): 363–368.

Kvasnytskyi, V. et al. 2020. Designing brazing filler metal for heat-resistant alloys based on ni al intermetallide. *Eastern-European Journal of Enterprise Technologies*. 6(12): 6–19.

Korzhyk, V. et al. 2021. Comparing Features in metallurgical interaction when applying different techniques of arc and plasma surfacing of steel wire on titanium. *Eastern-European Journal of Enterprise Technologies*. 4(112): 6–17.

Nguyen, D.H. & Widrow, B. 1990. Neural networks for self-learning control systems. *IEEE Control Systems Magazine*. 10(3): 18–23.

Welding and Related Technologies – Krivtsun et al. (Eds)

Peculiarities of the plasma-arc atomization of solid and flux-cored wires

I.V. Krivtsun & V.M. Korzhyk
China-Ukraine Institute of Welding, Guangdong Academy of Sciences, Guangzhou, China
E.O. Paton Electric Welding Institute, National Academy of Sciences of Ukraine, Kyiv, Ukraine

S. Gao*
China-Ukraine Institute of Welding, Guangdong Academy of Sciences, Guangzhou, China

D.V. Strohonov & O.M. Burlachenko
E.O. Paton Electric Welding Institute, National Academy of Sciences of Ukraine, Kyiv, Ukraine

X. Wang
China-Ukraine Institute of Welding, Guangdong Academy of Sciences, Guangzhou, China

O.I. Demianov
E.O. Paton Electric Welding Institute, National Academy of Sciences of Ukraine, Kyiv, Ukraine

ABSTRACT: Theoretical and experimental studies of forming primary drops of liquid metal and their subsequent fragmentation during plasma-arc wire atomization.

1 INTRODUCTION

It is known that the application of the technology of plasma-arc spheroidization of wire and rod materials (solid and flux-cored wires, flexible cords, rods, etc.) makes it possible to obtain high-quality spherical powders, the technological properties of which are at the level of powders obtained by the well-known industrial technologies of plasma rotating electrode process (PREP) and gas atomization (GA) of the melt [1,2]. At the same time, compared to the specified processes of obtaining spherical powders, the application of the technology of plasma-arc spheroidization of wire materials allows to significantly increase the output of the fine fraction <63 μm (50% wt. or more), which is in significant demand in the powder bed fusion (PBF) additive technologies, to improve the technical and economic characteristics of the process, which will further significantly reduce the cost of producing 1 kg of spherical powder [3-5].

For the specified technology of plasma-arc spheroidization, the study of the mechanisms of melting, gas-dynamic and thermal interaction of the melt formed during the melting of dispersed material with the plasma jet, the mechanisms of fragmentation of liquid metal drops are of significant scientific and practical interest. These issues still remain practically unstudied, which is absolutely necessary for the further industrial application of the technology of high-quality spherical powders production [6-9].

*corresponding author: meshiyigao@163.com

DOI: 10.1201/9781003518518-16

2 OBJECTIVES

The aim of the work is a theoretical and experimental study of the mechanisms and features of the processes of melting and fragmentation of the melt formed on the ends of solid and flux-cored current-carrying wires under various technological parameters of the plasma-arc spheroidization process, where to achieve the aim, the following tasks will be performed:

1. Mathematical modeling and analysis of calculated data on the influence of parameters of the plasma arc spheroidization mode of solid and flux-cored wires on the development of melt dispersion processes formed at the ends of the atomized wires and study of the patterns of fragmentation of the initial droplets in the plasma jet depending on the atomization distance.
2. Study of the mechanisms of fragmentation of the dispersed material of solid and flux-cored wires in the plasma jet using experimental methods of determining the particle size distribution of the powder obtained at different modes of the plasma-arc spheroidization process.

3 RESEARCH AND DISCUSSION

Calculations of the values of the diameters and speeds of movement of the liquid particle formed during the dispersion of the wire were carried out using a mathematical model of the jet flow of the molten metal of the wire and the formation of liquid metal droplets in the accompanying high-speed gas flow [9]. As can be seen from the presented calculation data, during plasma-arc spheroidization of a solid stainless steel wire AISI 316L with a diameter of 1.6 mm (Figure 1) and a flux-cored wire of the Fe-Al system with a diameter of 1.8 mm (Figure 1) multiple crushing is carried out particles and their fragments moving in the plasma jet. In particular, at different modes of operation of the plasma torch, initial drops are formed on the end of the sprayed solid AISI 316L wire, the average diameter of which lies in the range of 670…755 μm, and fragments (particles) are formed at a distance of 150 mm as a result of repeated crushing with an average diameter of 100…115 μm, while for the flux-cored wire of the Fe-Al system, the initial droplet diameter is 730…895 μm with an average diameter of fragments of 115-145 μm, respectively.

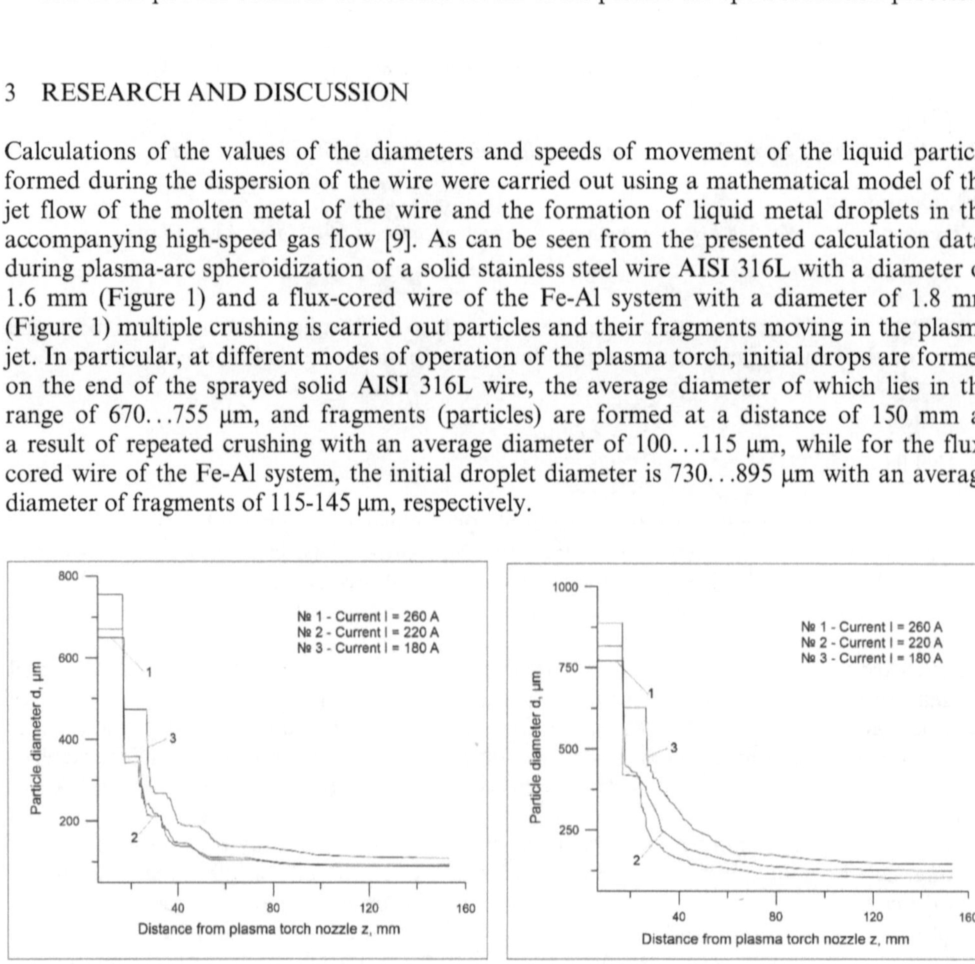

Figure 1. Change in the average diameter of the particles (d) formed on the end of atomized solid AISI 316L (left) and flux-cored wire of the-Fe Al system (right) in the process of plasma-arc spheroidization at different atomization modes, depending on the distance from the end of the plasma torch (z).

Based on the analysis of the results of calculations using the results of works [10-13], it is possible to distinguish the two most characteristic areas of crushing of liquid particles and their fragments. In the first section (near the section of the plasma torch nozzle), with a length of about

60 mm, which is characterized by high values of the speed and temperature of the plasma in the considered modes of plasma-arc spheroidization, the most intense crushing of particles is carried out. In this area, the We number for the particles significantly exceeds the critical value of the Weber number, which leads to their intense crushing. In the second section, due to the reduction (as a result of crushing) of the particle size, as well as the decrease in the relative velocity of the plasma and particles as they move away from the plasma torch nozzle section, the conditions of their flow change. At the same time, the molten particles experience a gradual increase in the force of aerodynamic resistance due to a change in their shape, which, after a certain time, can lead to their disintegration. Crushing of particles practically stops at a distance of about 120 mm from the cut of the plasma torch nozzle, because under the appropriate flow conditions, the time of development of the processes of deformation and destruction of particles becomes comparable. Analyzing the calculated and experimental data on the particles size distribution (PSD) of the powder obtained during plasma-arc spheroidization of the investigated solid and flux-cored wires (Figure 2), it can be seen that in both cases the main fraction is the fraction of 5-300 μm, which makes up to 97% wt., the yield of particles of the 300-400 μm fraction is at a rather low level and does not exceed 3% wt. When considering the PSD of the atomized particles in more detail, it should be noted that the error between theoretical and experimental data for the fraction 5-100 μm is no more than 16%, for fractions 100-200 μm no more than 7%, respectively. For the 200-300 μm fraction, the error in the results increases significantly and is 32%, and for the 300-400 μm fraction, it is 62%, respectively. The above data show that for the calculation of the main fraction, the error between theoretical and experimental results does not exceed 32%, which makes it possible to predict the powder PSD with a fairly good accuracy. It should also be noted that the share of the coarse fraction of 300-400 μm in the total volume is insignificant, so it does not have a significant effect on the overall change in PSD.

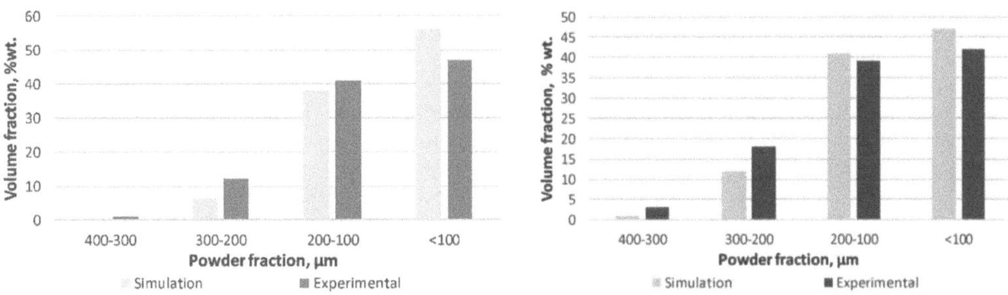

Figure 2. Histograms of the PSD of the powder obtained during plasma-arc spheroidization of solid stainless steel AISI 316L wire (left) and flux-cored wire of the Fe-Al system (right) at a current of 260 A.

4 CONCLUSIONS

1. A mathematical model was developed that describes the processes of thermal and dynamic interaction of the plasma jet with the melt formed at the ends of solid and flux-cored wires during plasma-arc spheroidization. This model makes it possible to determine the PSD of droplets that are formed as a result of fragmentation of the melt at the ends of solid and flux-cored wires due to the aerodynamic effect of the plasma jet for various parameters of the plasma arc spheroidization process mode.
2. Experimental verification of the results of mathematical modeling of the sizes of fragmented particles showed that in both cases of atomization both solid stainless steel AISI 316L wire and flux-cored wire of the Fe-Al system, the main fraction is the fraction of 5-300 μm, which is 97% wt. The error between the theoretical and experimental data for calculating the main fraction does not exceed 16-32%, which shows their quite satisfactory correspondence, which makes it possible to predict the PSD of the powder for determining the optimal modes of plasma arc spheroidization of a wide range of wire materials.

ACKNOWLEDGEMENTS

The work was funded by the Strategic project of the Academy of Sciences of Guangdong Province, (GDAS'Project of Science and Technology Development, 2020GDASYL-20200301001), China.

REFERENCES

1. Chen, G., Zhao, S., Tan, P., Wang, J., Xiang C.S., Tong, H.P. 2018. A comparative study of Ti–6Al–4V powders for additive manufacturing by gas atomization, plasma rotating electrode process and plasma atomization. *Powder Technology* (333): 38–46. (https://doi.org/10.1016/j.powtec.2018.04.013).
2. Sun, P., Fang, Z., Zhang, Y., Xia, Y. 2017. Review of the methods for the production of spherical Ti and Ti alloy powder. *JOM* (69): 1853–1860. (https://doi.org/10.1007/s11837-017-2513-5).
3. Zhang, W., Xu, Y., Shi, Y, Gu, Y., Volodymyr, K. 2022. Intergranular corrosion characteristics of high-efficiency wire laser additive manufactured Inconel 625 alloys. *Corrosion Science* (205): 110422. (https://doi.org/10.1016/j.corsci.2022.110422).
4. Gu, Y., Xu, Y., Shi, Y., Feng, C., Volodymyr, K. 2022. Corrosion resistance of 316 stainless steel in a simulated pressurized water reactor improved by laser cladding with chromium. *Surface and Coatings Technology* (441): 128534. (https://doi.org/10.1016/j.surfcoat.2022.128534).
5. Han, J., Shi, Y., Zhang, G., Volodymyr, K., Le, W.Y. 2022. Minimizing defects and controlling the morphology of laser welded aluminum alloys using power modulation-based laser beam oscillation. *Journal of Manufacturing Processes* (83): 49–59. (https://doi.org /10.1016/j.jmapro.2022.08.031).
6. Kharlamov, M.Yu., Krivtsun, I.V., Korzhyk, V.N. 2014. Dynamic Model of the Wire Dispersion Process in Plasma-Arc Spraying. *Journal of Thermal Spray Technology* (23): 420–430. (https://doi.org /10.1007/s11666-013-0027-4).
7. Prokopov, V.G., Fialko, N.M., Sherenkovskaya, G.P., Yurchuk, Y.L., Borisov, Yu.S., Murashov, A.P., Korzhik V.N. 1993. Effect of coating porosity on the process of heat transfer with gas-thermal deposition. *Powder Metallurgy and Metal Ceramics* (32): 22–26. (https://doi.org/10.1007 /BF00560034).
8. Korzhik, V.N. 1992. Theoretical analysis of the conditions required for rendering metallic alloys amorphous during gas-thermal spraying. III. Transformations in the amorphous layer during the growth process of the coating. *Powder Metallurgy and Metal Ceramics* (31): 943–948. (https://doi.org /10.1007/BF00797621).
9. Hryhorenko, G.M., Adeeva, L.I., Tunik, A. Yu., Karpets, M. V., Korzhyk, V. N. Kindrachuk, M. V., Tisov, O.V. 2020. Formation of microstructure of plasma-arc coatings obtained using powder wires with steel skin and B_4C + (Cr, Fe)$_7C_3$ + Al filler. *Metallofizika i Noveishie Tekhnologii* (42): 1265–1282. (https://doi.org/10.15407/mfint.42.09.1265).
10. Remesh, K., Yu, S.C.M., Ng, H.W., Berndt, C.C. 2003. Computational study and experimental comparison of the in-flight particle behavior for an external injection plasma spray process. *Journal of Thermal Spray Technology* (12): 508–522. (https://doi.org/10.1361/105996303772082251).
11. Li, H.P., Chen, X. 2001. Three-dimensional simulation of a plasma jet with transverse particle and carrier gas injection. *Thin Solid Films* (390): 175–180. (https://doi.org/10.1016/S0040-6090(01)00915-4).
12. Lasheras, J.C., Eastwood, C., Martinez-Bazán, C., Montañés, J.L. 2002. A review of statistical models for the break-up of an immiscible fluid immersed into a fully developed turbulent flow. *International Journal of Multiphase Flow* (28): 247–278. (https://doi.org/10.1016/S0301-9322(01)00046-5).
13. Marchetti, J.M., Patruno, L.E., Jakobsen H.A., Svendsen, H.F. 2010. Mathematical framework for higher order breakage scenarios. *Chemical Engineering Science* (65) 5881–5886. (https://doi.org/10 .1016/j.ces.2010.08.008).

Welding and Related Technologies – Krivtsun et al. (Eds)

Changes in the structure and properties of high chromium cast iron soldered with carbide encapsulation in a charge of powdered wire

V. Peremitko, V. Sukhomlyn & I. Kolomoyets
Dniprovsky State Technical University, Kamianske, Ukraine

A. Golyakevych
TM Weltech, Kyiv, Ukraine

A. Yevdokymov
LLC Steel Work, Kryvyi Rih, Ukraine

G. Wilhelm
München University of Applied Sciences, Munich, Germany

ABSTRACT: Experiments have demonstrated the effectiveness of introducing chromium carbide as dispersed particles, coated with a protective plastic layer, during arc hardfacing. The VELTEK-N630-O powder wire was used, and metallographic analysis revealed differences in the shape, amount, and distribution of carbides in deposited layers similar to high-chromium white cast iron. An increased proportion of carbides in the upper part of the layers and fewer cracks should improve wear resistance. Microhardness measurements confirmed a lower initial hardness (HB 483 vs. HB 510 for standard wire), due to the addition of plastic. However, wear tests in a "metal-metal" scheme showed reduced mass loss in samples welded with the experimental wire: 0.158 g versus 0.208 g after 12 hours of rolling friction under a load of 1......1.5 kN

1 INTRODUCTION

The service life of machines in mineral extraction, metallurgy, and construction is often limited by wear. To extend their lifespan, materials with enhanced wear resistance are used, along with surface strengthening techniques like chemical and thermal treatment, spraying, and electric arc hardfacing (Cherneta et al. 2019, Hruzevich & Shvets 2021). White cast iron, alloyed with elements such as chromium, titanium, molybdenum, and vanadium, has proven effective. To overcome martensite's brittleness, increasing bainite and austenite proportions is recommended. In white cast irons, refining the carbide phase, especially with chromium, improves wear resistance (Kindrachuk et al. 2016, Brykov 2020).

For long-term use, contact surfaces must balance strength and toughness. Alloy performance can be enhanced through modification and microalloying (Babinets & Ryabtsev 2021, Voronchuk et al. 2022) in electric arc hardfacing, although some loss of charge components occurs in the arc gap. This study examines the effect of coating modifier particles in the powder wire charge with plastic on the structure and properties of white cast iron layers applied by arc deposition. Adding plastic to the powder tape slows crystallization, enhances carbon diffusion to chromium carbides, and depletes nearby austenite to cementite (Peremitko et al. 2024).

2 MATERIALS AND METHODS

For the experiments, 09G2S low-alloy steel samples (150x250x8 mm) were used, onto which powder-coated wires similar in composition to VELTEK-N630-O (2.4 mm diameter) were

DOI: 10.1201/9781003518518-17

deposited to create a 4-5 mm wear-resistant layer. The wire contained 5.4% C, 1.2% Mn, and 0.016% S, with the main difference being the pre-coating of chromium carbide particles with plastic. Two experimental surfacing materials were prepared using VELTEK-N630-O wire: one with chromium carbide coated in Unikast 10/14 polyurethane, and the other in fluoroplastic (50 g per 1360 g of chromium carbide). Plastic was sprayed onto 0.3-0.4 mm carbide particles from 200-250 mm, dried at 120°C for 20 hours, and sieved through a 400μm sieve.

For comparison, standard VELTEK-N620 and VELTEK-N630-O wires without plastic coatings were used. Electric arc surfacing was performed automatically (I_h = 220-250 A, U_a = 25-28 V, v_h = 220 mm/min) in two passes on the "Krystal PNP-2.5" installation. Spectral analysis of the deposited metal's composition was done using an optical emission spectrometer (SPECTRO MAXx), and metallographic studies were conducted via electron microscopy (REM-106i). Wear resistance was tested on the MI-1M friction machine with a 100 N load under rolling, sliding, and rolling with sliding conditions.

3 EXPERIMENTAL RESULTS AND DISCUSSION

Arc burning with the experimental wires remained stable, with electrode metal transfer varying from large drops to a jet, and minimal surface slag formation, while the rollers were uniform. A comparison of the chemical composition of the deposited metal using standard and experimental powder-coated wires (Table 1) showed notable changes.

Table 1. Chemical analysis of the deposited layer.

Sample serial number	Material's brand	C	Mn	Si	Cr	Ni
1.	H-630 without plastic	3.85	0.23	1.36	22.03	0.03
2.	H-630 with Unikast 10/14	4.93	0.28	1.37	25.95	0.03
3.	H-630 with fluoroplastic	4.95	4.95	1.51	24.52	0.03

In the case of encapsulation of chromium carbide particles (samples 2 and 3), an increase in the presence of carbon in the heated metal is recorded. Moreover, the use of polyurethane (Unikast 10/14) caused a more significant jump in the values of carbon and chromium content.

The microstructure of the metal welded with standard flux-cored wire VELTEK-N-630 (without the addition of plastic) has a ladyburite matrix, in which excess carbides are fixed in the form of lamellar discharges (sample 1, Figure 1,a). Moreover, in the fusion zone, the sizes of carbide deposits are smaller.

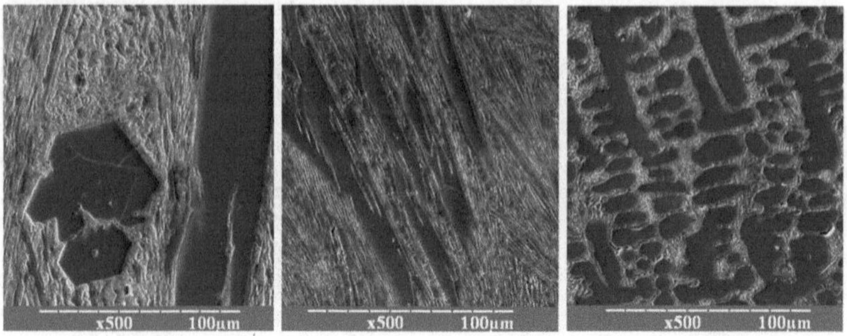

a b c

Figure 1. Differences between structure of layers, flooded by standart (sample № 1, a) and experimental wires (sample № 2 – b; sample № 3 – c).

When using flux-cored wire coated with chromium carbide particles along with Unikast 10/14 polyurethane during hardfacing, the metal forms a super-eutectic white cast iron structure (sample 2, Figure 1b), with more dispersed carbide deposits. Above the fusion line, fine dendrites and a mixture of ledeburite, cementite, and pearlite are observed, which is attributed to changes in the carbon and chromium ratio. With the addition of powdered fluoroplastic (sample 3, Figure 1c), further refinement of carbide particles is noted, even after applying three layers by arc hardfacing. In both series 2 and 3, lamellar carbides predominantly align perpendicular to the fusion line, except for clusters near the surface with varying orientations.

To find out the differences in the properties of the deposited layers, an additional study of their resistance to dry wear using the "metal by metal" scheme was conducted. A comparison of mass loss and hardness changes due to wear is shown in Tables 2-3.

Table 2. Conditions and results of research of wear-resistance of hardfaced layers with a load of 1 kN.

The composition of hardfacing wire's charge	Test time, hours	Mass loss at 1 kN	Hardness, HB at 1 kN
Veltek H-630, sample No. 1	2	0.12060	431
	4	0.06830	424
	6	0.08845	492
Veltek H-630+Unikast, sample No. 2	2	0.09604	347
	4	0.10826	404
	6	0.34490	447

Table 3. Conditions and results of research of wear-resistance of hardfaced layers with a load of 1.5 kN.

The composition of hardfacing wire's charge	Test time, hours	Mass loss at 1 kN	Hardness, HB at 1 kN
Veltek H-630, sample No. 1	2	0.46435	628
	4	0.22015	614
	6	0.34015	600
Veltek H-630+Unikast, sample No. 2	2	0.39140	587
	4	0.43850	620
	6	0.48850	625

During the testing of sample No. 1 at the run-in stage, a decrease in hardness and mass loss was observed, attributed to the eutectic cast iron (3.85% C) structure formed by Veltek H-630 wire. The increased load led to higher temperatures in the contact zone, enhancing plasticity in the near-surface layer. The introduction of plastic into the charge reduced the initial hardness of the applied layer by 5.3% compared to standard wire. Sample No. 2 exhibited deformation hardening throughout the wear period, with a simultaneous increase in hardness and mass loss, likely due to high carbon and silicon content. Wear testing showed that the experimental wire reduced mass loss to 0.158 g compared to 0.208 g over 12 hours under 1 kN load.

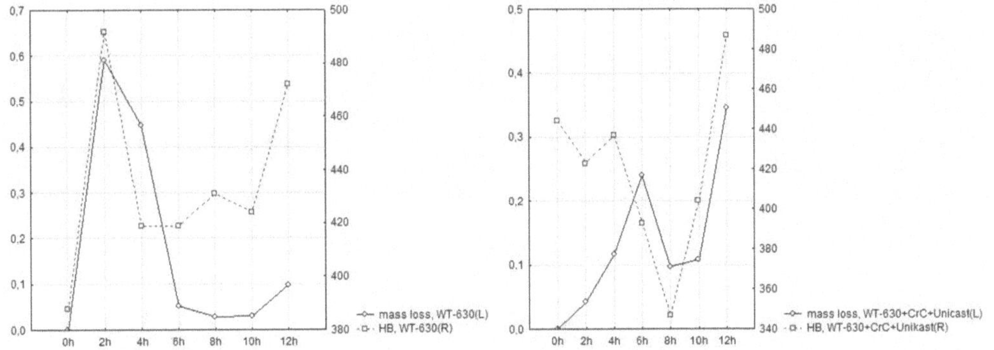

a b

Figure 2. Change in hardness and mass loss during 12 hours of research (samples welded with a standard wire, a; samples welded with an experimental wire, b).

In comparison with the case of the test at 1 kN, the predicted intensification of the destruction is observed due to the increase in the load on the contact surface, which is confirmed by the increase in mass loss by an average of 7 times (Figure 3a,b). Under conditions of increased loads, a significant role is played by the deformation hardening of the samples and, as a result, their embrittlement. The difference between the wear process of the samples lies in the duration of the strain hardening process. Samples with lower silicon content show a longer onset of embrittlement.

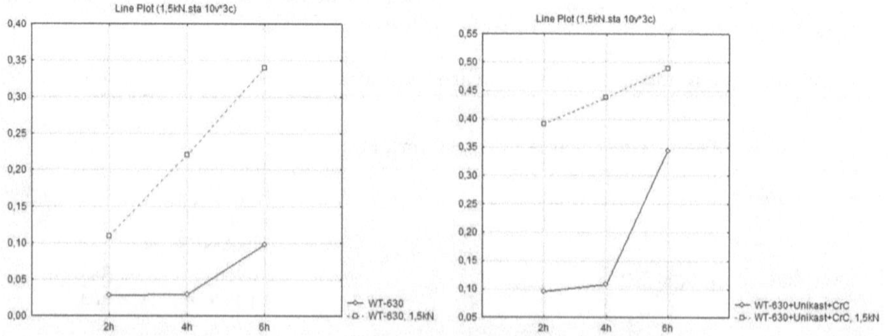

a b

Figure 3. Comparison of the change in mass loss during 6 hours of actual wear under a load of 1…1.5 kN (samples welded with a standard wire, a; samples welded with an experimental wire, b).

4 CONCLUSIONS

The effectiveness of the scheme for regulating the composition, structure, and properties of the deposited metal due to the preliminary coating of the functional component of the flux-cored wire charge with plastic was experimentally confirmed.

It has been established that the use of plastic during the formation of the charge of the hard-facing wire contributes to the grinding of structural components, primarily carbides, ensures the saturation of the heated metal with carbon and the formation of a structure favorable for wear resistance.

The addition of plastic to the charge caused a decrease in the initial hardness of the metal of the applied layer by 5.3% compared to the case of using a standard wire. A metal-to-metal wear test recorded a 1.5-fold reduction in mass loss of the samples welded with the test wire in 12 hours. rolling friction under a load of 1 kN.

REFERENCES

Babinets, A.A. & Ryabtsev, I.O. 2021. Influence of modification and microalloying on deposited metal structure and properties (Review). *The Paton Welding Journal* 10: 3–11.

Brykov M.M. 2020. Wear resistance of steels under mechanical wear conditions. *Metalozn. obrobka metaliv* 26: 54–62.

Cherneta O.G., Sukhomlyn V.I. & Kalinichenko S.V. 2019. Study of the restored surface layer of parts made of steel 45 with a strengthening treatment for wear resistance. *Mathematical modeling* 1(40): 194–201.

Hruzevich A.V. & Shvets V.V. 2021. Increasing the reliability of power plant equipment by means of electric arc spraying. *The Paton Welding Journal* 1: 40–48.

Kindrachuk M.V., Kutsova V.Z., Kovzel M.A. & Tisov O.V. 2016. Modern functional materials with a bainite nanostructured matrix and increased tribological properties. *Friction and Wear Problems* 70: 112–130.

Peremitko V., Golyakevich A., Sukhomlyn V. & Artemchuck V. 2024. Einfluss des Verfahrens zur Zugabe von Karbidbildnern auf die Struktur und die Eigenschaften aufgeschweißter Schichten.. *Schweissen und Schneiden* 1-2: 40–46.

Voronchuk O.P., Zhudra O.P. & Kaida T.V. 2022. The influence of the composition of the components of the charge of powder ribbons of the C–Fe–Cr–Nb alloying system on the chemical composition and structure of the deposited metal *The Paton Welding Journal* 2: 29–34.

Welding and Related Technologies – Krivtsun et al. (Eds)
© 2025 The Author(s), ISBN 978-1-032-85176-1
Open Access: www.taylorfrancis.com, CC BY-NC-ND 4.0 license

The influence of the welding thermal cycles on the distribution of alloying elements in the HAZ structure of duplex steel 2205

S.Yu. Maksymov & G.V. Fadeeva
E.O. Paton Electric Welding Institute, National Academy of Sciences of Ukraine, Kyiv, Ukraine

Chuanbao Jia
Key Laboratory of Liquid-Solid Structural Evolution and Processing of Materials of Ministry of Education, Institute of Materials Joining, Shandong University, Jinan, PR China

D.V. Vasiliev & A.A. Radzievska
E.O. Paton Electric Welding Institute, National Academy of Sciences of Ukraine, Kyiv, Ukraine

ABSTRACT: The article presents the results of determining the distribution of alloying elements in δ-ferrite and austenite in the microstructure of the HAZ simulated using the Gleeble-3800 device for welding thermal cycles that are inherent in wet underwater welding and air welding. It is shown that with an increase in the cooling rate, the difference in the redistribution of the elements Mo and Cr, as well as N and Ni between the ferrite and austenite phases decreases. The highest content of these alloying elements is observed at the boundary of the δ/γ phases.

1 INTRODUCTION

In duplex stainless steels (DSS), the optimal combination of high strength characteristics and corrosion resistance, primarily against pitting corrosion, as well as stress corrosion cracking, is ensured by the ratio of ferrite and austenite phases in the metal structure as 1:1 (Labanowski, J. et al. 2011). Under the influence of the welding thermal cycle (WTC), due to the effect of high heating and cooling rates, the phase balance is disturbed, which in turn affects the mechanical properties and corrosion resistance of welded joints (Maksimov S.Yu. et al. 2023). The study of the distribution of alloying elements Cr, Mo, Ni in δ-ferrite and austenite in duplex steels as received (Vicente A.A. et al. 2020) to predict and determine the pitting corrosion resistance of steels showed that due to the difference in the chemical composition of the ferrite and austenite phases, the PREN value also differs. In (Zhang Y., Liu Z. et al. 2021), the effect of nitrogen on the microstructure of the weld metal and HAZ, the distribution of alloying elements in the austenite and ferrite phases and the corrosion resistance of welded joints in GTAW were investigated. Welding was carried out in a shielding gas of 100% Ar (D1) and with the addition of nitrogen Ar + 2% N_2 (D2). Table 1 shows the content of alloying elements in the ferrite and austenite phases in the weld metal and HAZ and the PREN value in D1 and D2 welded joints.

The study of nitrogen diffusion in austenite and ferrite of 25Cr super duplex steel as a result of gas nitriding experiments (Zhang Y., Qiu, Z. et al. 2020) showed that nitrogen diffused much faster in the austenite phase than in the ferrite phase. In the work (Bae S.H. et al. 2013), the corrosion resistance, phase composition change and distribution of alloying elements in the austenite and ferrite phases, as well as nitrogen, depending on the molybdenum content from 0.57 to 3.52 wt.% in the weld metal were studied. It is shown that the localisation of pitting corrosion occurs selectively with the point of pitting initiation in the chromium and molybdenum-depleted zone at the grain boundary between γ and δ-ferrite. As can be seen from the examples given, the diffusion of alloying elements in austenite and ferrite affects the formation of the microstructure and the distribution of chemical elements in the material, which affects the mechanical

DOI: 10.1201/9781003518518-18

properties, corrosion resistance and other characteristics. Optimum distribution and concentration of alloying elements in each phase help to increase the strength, crack resistance and corrosion resistance of duplex steels. Local determination of the nitrogen content in austenite of various morphologies, in particular grain boundary (GBA), Widmanstatten (WA) and intragranular (IGA) in duplex steels is especially important for understanding the structural features of these materials and their effect on properties. Local analysis of the nitrogen content in various austenite morphologies allows for a better understanding of the formation mechanisms of structural phase transformations and the impact of nitrogen on this process.

Table 1. Mass percentage of chemical elements in ferrite and austenite phases in GTAW welding with 100% Ar (D1) and 98% Ar + 2% N_2 (D2) shielding gas.

Region	Phases	Element, mass.%				PREN
		Cr	Mo	Ni	N	
Base metal	Ferrite	22.68	3.73	4.51	0.121	36.92
	Austenite	21.86	3.55	7.13	0.144	35.90
	Ferrite/Austenite phase boundary	21.37	3.17	3.65	0.128	33.88
Weld metal (D1)	Ferrite	21.74	3.52	8.02	0.082	34.67
	Austenite	21.04	2.37	8.65	0.120	30.78
	Ferrite/Austenite phase boundary	20.50	3.12	5.45	0.102	32.43
HAZ (D1)	Ferrite	20.96	3.88	5.45	0.115	35.60
	Austenite	20.84	2.57	5.46	0.142	31.59
	Ferrite/Austenite phase boundary	20.19	2.64	4.30	0.123	30.87
Weld metal (D2)	Ferrite	22.34	3.50	8.90	0.188	36.90
	Austenite	20.49	2.56	10.19	0.195	32.06
	Ferrite/Austenite phase boundary	18.01	2.86	7.37	0.200	30.65
HAZ (D2)	Ferrite	20.15	3.37	5.11	0.204	34.54
	Austenite	19.06	2.84	6.56	0.219	31.94
	Ferrite/Austenite phase boundary	21.37	3.17	3.65	0.128	33,88

In the literature, there are almost no studies on the distribution of alloying elements and nitrogen in the austenite and ferrite phases at cooling rates different from the usual ones, for example, when welding in an aqueous medium. Therefore, the study aimed to determine the distribution of alloying elements and nitrogen in ferrite and austenite of different morphologies, as well as on the γ/δ boundary in HAZ depending on the cooling rate during welding.

2 METHODOLOGY AND RESEARCH METHODS

Modelling of microstructures was performed on a Gleeble-3800 installation according to the thermal procedure: heating at a rate of 100 °C/s to a temperature of T = 1300 °C, holding at a peak temperature of τ = 2s, cooling of samples according to WTC in air welding and wet underwater welding (Hasui, A. et al. 1980)

After thermal modelling, samples were cut out for further studies using optical microscopy (OM) and analytical scanning electron microscopy (SEM). The austenite and ferrite content in the simulated HAZ samples was determined using MIPAR image analysis software. Optical metallography was performed on Versamet-2 (USA) and Neophot-32 (Germany) microscopes. Scanning electron microscopy (SEM) and energy-dispersive spectroscopy (EDC) were used to determine the distribution of alloying elements and nitrogen using a Tescan Mira 3 LMU scanning electron microscope (Czech Republic) with an Oxford Instrument X-MAX 80 mm^2 characteristic X-ray detector. The equivalent number of resistance to pitting corrosion was determined by the formula: PREN = wt.% Cr + 3.3 wt.% Mo + 16 wt.% N.

3 EXPERIMENTAL RESULTS AND DISCUSSION

The distribution of alloying elements and nitrogen in the ferrite and austenite phases in the HAZ microstructure of duplex stainless steel 2205 (EN 1.4462) was investigated. The cooling rate was different (temperature range T = 1300 – 800 °C) and the values were as follows: $W_{13/8}$ = 81.70 °C/s, $W_{13/8}$ = 165.85 °C/s and $W_{13/8}$ = 320.51 °C/s. The results of determining the distribution of alloying elements in the HAZ microstructure between the ferrite and austenite phases and at the δ/γ boundary are shown in Table 2.

Table 2. The distribution of alloying elements between the ferrite and austenite phases in the HTHAZ microstructure (zone I – large grain zone, temperature range T = 1300-800 °C) depends on the cooling rate.

№ sample, $W_{13/8}$, °C/c, τ,	Area of analysis	The name of the phases	Content of elements, mass. %				PREN
			N	Cr	Ni	Mo	
Certification data as delivered	BM 2205		0.152	22.146	4.931	2.557	33.00
1 – air, the middle of weld, $W_{13/8}$ = 8.21, τ = 60.88	1	Ferrite	0.00	21.85	5.38	3.19	32.38
	2	Austenite GBA	0.50	21.59	7.47	2.65	38.34
	3	Austenite GBA	0.40	21.80	4.13	2.88	37.70
	4	Ferrite	0.00	21.73	6.11	3.16	32.16
	5	Austenite WA	0.73	21.46	4.86	2.47	41.27
	6	Ferrite	0.00	22.69	4.33	3.26	33.45
	7	Austenite rhombus	0.34	21.17	6.87	2.12	33.60
	8	Austenite rhombus	0.40	22.08	5.08	2.56	36.93
2 - water, the middle of the weld, $W_{13/8}$ = 81.70, τ = 6.12	1	Ferrite	0.25	21.21	6.83	3.49	36.72
	2	Near the border δ/γ	0.27	22.20	7.35	3.47	37.97
	3	Ferrite	0.00	23.76	3.39	3.12	34.06
	4	Border δ/γ	0.98	21.76	6.48	3.41	48.69
	5	Reformed ferrite	0.65	19.34	7.07	2.07	36.57
	6	Austenite	0.83	20.16	4.50	2.50	41.69
3 – water, the beginning of the weld, $W_{13/8}$ = 165.85, τ = 2.94	1	Ferrite	0.00	21.88	4.17	2.57	30.36
	2	Ferrite	0.00	21.00	5.90	2.82	30.30
	3	Near the border δ/γ	0.29	21.35	7.22	2.91	35.59
	4	Near the border δ/γ	0.64	21.04	5.38	2,16	38.40
	5	Austenite	0.73	21.49	5.70	2.62	41.81
4 – water, crater, $W_{13/8}$ = 320.51, τ = 1.56	1	Ferrite	0,00	23,33	6,97	3,68	35,47
	2	Reformed ferrite	0,37	22,42	2,44	2,50	36,39
	3	Austenite	0,62	20,75	4,92	2,08	37,53
	4	Austenite	0,53	21,15	7,16	2,14	36,69
	5	Reformed ferrite	0,27	22,12	4,73	2,90	36,01
	6	Ferrite	0,00	22,07	5,38	2,42	30,06

The analysis of the results presented in Table 2 shows that PREN in sample 1 in the austenite phase varies from 33.60 to 41.29, and the range of PREN in δ-ferrite is from 32.16 to 33.45. That is, min PREN in austenite equals max PREN in δ-ferrite. In sample 2, PREN in austenite varies from 41.69 to 48.69, and in δ-ferrite from 34.06 to 37.97. The PREN value in sample 2 is higher in austenite and δ-ferrite than in sample 1. In sample 3, PREN in austenite is 41.81, and PREN in δ-ferrite is from 30.30 to 38.40. In sample 4, PREN in austenite varies from 36.39 to 37.53, and in δ-ferrite from 30.06 to 35.47. PREN of the austenite and δ-ferrite phases already differ to a lesser extent than in the previous samples. The highest PREN values are at the δ/γ boundary (sample 2) and are equal to 48.69, while the N content is 0.98 wt.%, as well as in the austenite phase (sample 1) and are 41.27, while the N content is 0.73 wt.%, and in sample 3 PREN of austenite is 41.81, the nitrogen content is 0.73 wt.%. The PREN value of δ-ferrite in all samples is lower than in the austenite phase and ranges from 30.06 to 36.72. This emphasizes that it is precisely due to the diffusion of alloying elements, primarily nitrogen, that the difference in the PREN values of the phases is reduced since the time spent in the temperature range of phase transformations is diminished.

The highest microhardness value is possessed by austenite in HTHAZ, both Widmanstätten (WA) and grain boundary (GBA), HV 3300...5150 MPa. The microhardness of austenite decreases with the increase in the cooling rate from $W_{13/8}$ = 8.21 °C/s to $W_{13/8}$ = 320.51 °C/s and is in the range of HV 3360...4390 MPa. The microhardness of austenite in OM is mainly HV 2970...3090 MPa, sometimes reaching HV 3300...3570 MPa. The change in the microhardness of austenite may indicate that the main element influencing the microhardness of austenite is nitrogen. The microhardness depends on its content in the austenite lattice. The nitrogen content in austenite ranges from 0.40 wt.% to 0.98 wt.%. The microhardness of ferrite does not change depending on the cooling rate and is equal to HV 2300...2900. The microhardness of ferrite in the base metal is almost the same HV 2300...2570 MPa.

The greatest variations from the average value in BM (certificate data) are observed for Mo, especially in its distribution in δ-fertite, namely, for sample 1 the range is from 3.16...3.26 wt.%; for sample 2 from 2.07...3.49 wt.%; for sample 3 – 2.57...2.82 wt.% and for sample 4 – 2.42...3.68 wt.%. The increase in the range in sample 4 can be explained by grain refinement at a high cooling rate. It is also for N, especially in austenite. For sample 1 the range is 0.34...0.73 wt.%; for sample 2 – 0.83...0.98 wt.% (at the δ/γ boundary); for sample 3 – 0.29...0.64 wt.% and for sample 4 – 0.53...0.62 wt.%.

4 CONCLUSIONS

1. The effect of cooling rate on the distribution of alloying elements in the ferrite and austenite phases in the simulated HAZ of 2205 duplex steel was studied under different welding thermal cycles using Gleeble-3800.
2. Depending on the cooling rate, namely on the time spent in the temperature range of phase transformations, a redistribution of alloying elements between the austenite and ferrite phases is observed, both ferrite-forming elements Mo and Cr and austenite-forming Ni and N. With an increase in the cooling rate to $W_{13/8}$ = 320.51 °C/s, the difference in the redistribution of alloying elements between the phases decreases.
3. Austenite of different morphology differs in the magnitude of microhardness values. A change in the microhardness of austenite of different morphology can result from the content of different amounts of alloying elements, primarily nitrogen in the austenite lattice. At the same time, the microhardness of ferrite does not change depending on the cooling rate. It remains almost constant, both in the HTHAZ (I and II zones) in the temperature range T = 1300-800 °C, and in the LTHAZ (III zone) in the temperature range T = 800-500 °C and is equal to the microhardness of ferrite in the base metal. The nitrogen content in austenite is from 0.40 wt.% to 0.98 wt.%.
4. A decrease in the microhardness of austenite in the HTHAZ (T = 1300 ... 800 °C) occurs with an increase in the cooling rate from $W_{13/8}$ = 8.21 °C/s to $W_{13/8}$ = 320.51 °C/s and is in the range of HV 3360 ... 4390 MPa. The microhardness of austenite in BM is mainly HV 2970...3090 MPa, sometimes reaching HV 3300...3570 MPa.

5. The change in the microhardness of austenite may indicate that the main element influencing the microhardness of austenite is nitrogen. The microhardness depends on its content in the austenite lattice. The highest microhardness value is for austenite in the HTHAZ, both Widmanstätten (WA) and grain boundary (GBA), HV 3300...5150 MPa.

REFERENCES

Bae S.H. & Lee H.W. 2013. Effect of Mo Contents on Corrosion Behaviors of Welded Duplex Stainless Steel. *Met. Mater. Int.*, *19, No. 3*, 563~569. *doi*: *10.1007/s12540-013-3026-6.*

Hasui, A. & Suga, Y. 1980. On cooling of Underwater Welds. *Trans. of the Japan Welding Society. 11, № 1.*

Labanowski, J. Et al. 2011. Underwater welding of Duplex Stainless Steel. *Solid State Phenomena*, *183*, 101-106.

Maksimov S.Yu. et al. 2023. Peculiarities of duplex stainless steels (DSS) welding during wet underwater welding compared with welding in air. Automatic welding. 10, 10–18.

Vicente A.A. et al. 2020. Study of the Distribution of Cr, Mo, Ni and N in δ Ferrite and Austenite in Duplex Stainless Steels. *Journal homepage*: https://saudijournals.com/sjeat.

Zhang Y., Liu Z. et al. 2021. Effects of nitrogen element on microstructure and pitting corrosion resistance of duplex stainless steel welded joints. Mater. Res. Express. 8. http://dx.doi.org/10.1088/2053-1591/abe42f.

Zhang Y., Qiu, Z. et al. 2020. Nitrogen diffusion behaviour in austenite and ferrite of 25Cr super-duplex stainless steel during gaseous nitrocarburizing. *Materials Characterization*, *159.*

Welding and Related Technologies – Krivtsun et al. (Eds)

Laser synthesis of CBN-contained abrasive layers

O.O. Goncharuk, O.D. Kaglyak, D.A. Lesyk, L.F. Golovko, S.M. Voloshko & O.A. Savchenko
Laser Systems and Advanced Technologies Department, National Technical University of Ukraine "Igor Sikorsky Kyiv Polytechnic Institute", Kyiv, Ukraine

ABSTRACT: Cubic boron nitride (CBN) is widely used for abrasive disk manufacturing for steel processing. In the given work some aspects related to the increase of manufacturing productivity for the abovementioned cutters are portrayed, as well as technological processes of laser synthesis for the manufacturing of abrasive layers. Results of laser irradiation absorbability by the CBN depending on its state or on its powder dispersion and surface roughness, heating regimes and initial temperature are presented in details.

1 INTRODUCTION

Research on the use of laser radiation for straightening abrasive tools was studied by a number of authors (Bing et al. 2022). A pulsed Nd: YAG laser was used for the ablation of grinding wheels based on Al_2O_3 and SiC ceramics, which proved the possibility of using a pulsed laser for straightening grinding wheels, and the irradiation of a grinding wheel on a ceramic base with a CBN CO_2 laser showed a high perspective of the chosen path. The results showed that the combination of laser editing and mechanical editing was effective.

Preliminary studies conducted by the authors showed the possibility of using the technology of laser synthesis of working layers of an abrasive tool based on cubic boron nitride and a metal binder. Technological parameters of irradiation were also established, which did not lead to thermal destruction of the abrasive grains.

Experiments were carried out using two industrial laser systems: ROFIN DY 044 based on Nd:YAG laser with diode pimping and "LATUS 31" based on CO_2 laser with longitudinal pumping of gas mixture. A powder feeding unit and laser scanning systems were used with each technological laser. Technological parameters for LATUS 31 (λ=10,6 microns) were secured in the following range: power density Wp=(0.3 - 6.37)x10^4 W/cm^2, processing time τ =0,15-0,45 s). For laser "DY 044" with wavelength λ=1,06 microns (power density Wp=(0.1-1.4)x10^4 W/cm^2, processing time τ =0,06-4,5 s). Laser processing was done in air and in argon atmosphere with gas flowrate 2-20 l/min.

The results of standard strength analysis of CBN grains that were directly irradiated by laser beam at different wavelengths and in different protective atmosphere are shown in Figure 1.

Cubic boron nitride (CBN) is widely used for abrasive disk manufacturing for steel processing. In the given work some aspects related to the increase of manufacturing productivity for the abovementioned cutters are portrayed, as well as technological processes of laser synthesis for the manufacturing of abrasive layers. Results of laser irradiation absorbability by the CBN depending on its state or on its powder dispersion and surface roughness, heating regimes and initial temperature are presented in details.

Results of strength measurements of CBN grains are in a good coincidence with the data from X-ray diffraction analysis – two highly intense reflections with interference indexes 111 and 220 (Goncharuk 2017). The first maximum with relative intensity of 100 % is related to atomic plates with distance d = 2,087 Å between the planes and was observed at angle 2θ = 43,360. This observed intensity is determined by the quantity of CBN phase in the powder

DOI: 10.1201/9781003518518-19

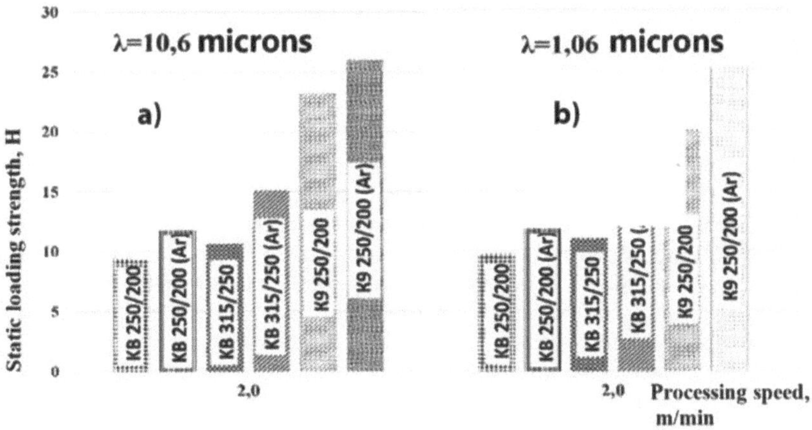

Figure 1. Strength of irradiated CBN grains in air and argon under static load: a) d=3 mm, P=800 W; b) d=3 mm, P=500 W.

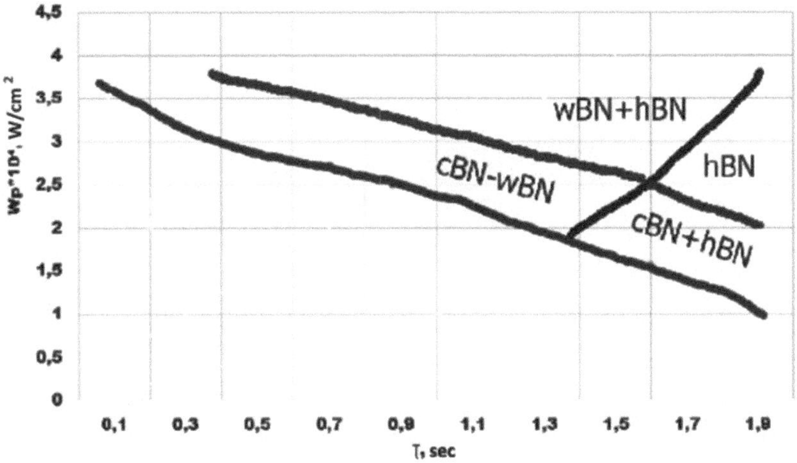

Figure 2. Dependence of the phase composition of the KV250/200 powder on energy parameters of laser radiation.

mixture. With the increase of laser beam power density and duration of laser beam irradiation the decrease of CBN phase is observed. In turn, the graphite phase increases and these changes are represented through the intensity lines of corresponding phases.

At indirect irradiation the processing regimes shift into the "harsh" zone with no significant loss in strength on CBN. It could be seen from the results of SEM analysis that the appearance of micro-cracks on grains and the initiation of cBN→hBN phase changes when sintering is done at scanning speed of 0,6-0,8 m/min, laser beam power 800 W and laser beam focal spot - 3 mm. It means that it is possible to use bonding materials with melting points higher than 1500°C. SEM analysis confirms the presence of the crystal-like structure of CBN and strength measurements also correlate with data from SEM analysis (Goncharuk et al. 2018).

Comparing the results of direct and indirect laser irradiation and its technological parameters on CBN grains, the geometrical characteristics and micro-hardness of sintered layers and distribution of bonding elements in them, there were established energy parameters and optimal areas of technological regimes of laser forming of tool composites Figure 3 and Figure 4. With these regimes it is possible sinter layers of a high quality using two types of lasers with wavelengths of 10,6 microns и 1,06 microns correspondingly.

When laser source with wavelength of λ=10,6 microns is used, it is recommended to use untreated heat resistant CBN grains (K9) with grain size of 160-315 microns in a junction with bonding materials that have a melting temperature no higher than 1100°C (Br010, Cu+Sn). If it is necessary to use more refractory bondings (Ni and Fe based) with the melting temperatures of 1400-1600°C then CBN grains with Ni coating in the best option. The area of appropriate technological parameters are as follows: Wp= $(0.5\text{-}2.0)\text{x}10^3$ W/cm^2 with irradiation time 0.23-3.0 s in the protective atmosphere (Argon, Nitrogen).

For laser sources with the wavelength λ=1,06 microns, due to high absorption, it is recommended to use K9 and KB CBN grains with grain size 160-315 microns that are vacuum-coated with metals (Cu, Ni, Ti). Their melting temperatures are higher than that of bonding material and it is possible to maintain strength characteristics of CBN grains after laser processing. The key technological parameters are: Wp= $(0.2\text{-}0.8)\text{x}10^3$ W/cm^2 with the irradiation time 0.1-1.5 s in the protective atmosphere. The presence of protective atmosphere significantly influences on the formation of sintered layer and this effect could be used for selective laser sorting and laser modification of CBN grains.

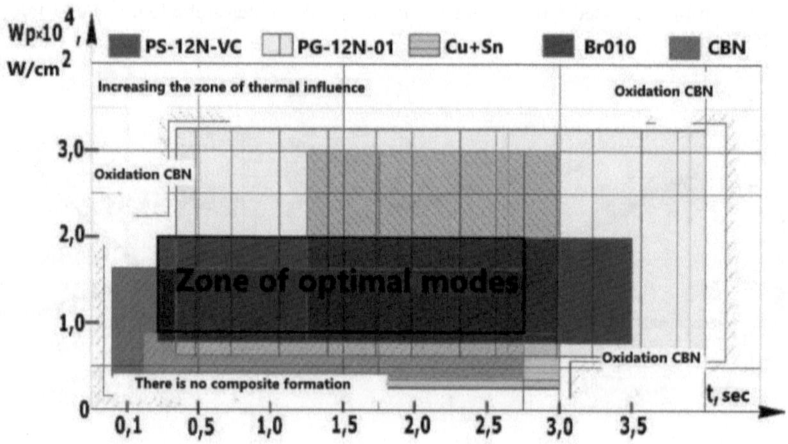

Figure 3. The range of optimal technological modes of laser sintering of CBN-containing composites by radiation with λ=10.6 μm.

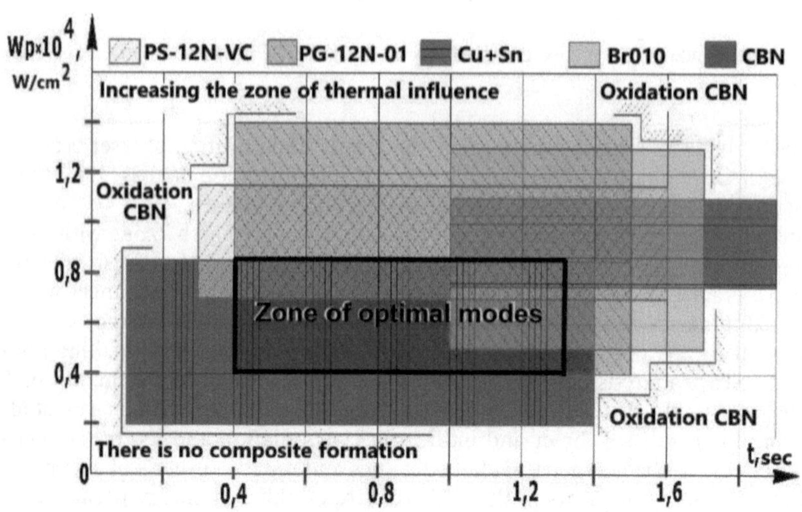

Figure 4. The range of optimal technological modes of laser sintering of CBN-containing composites by radiation with λ=1.06 μm.

Elements distribution map of sintered composition Br010+KV315/250 with CBN grains vacuum-coated with copper shows that there is an intensive intermixing of composite powder with partial immersion of CBN grains in bonding powder. This statement is supported by the presence of Cu layer on the grain surface with thickness less than 15 microns and Fe layer on the surface of abrasive that was migrated from the specimen surface during laser melting. The concentration of Cu on the boundary of grain-bonding powder confirms the functionality of the pre-coating of CBN grains. Coated layer was partially melted and dissolved in the composite volume and Cu was mainly concentrated close to CBN grains. In order to test recommended processing regimes several abrasive tools were manufactured.

In order to reduce the thermal load on the tool body and to extent the limits of powder types used for sintering it was decided to use the intermittent layer that is used as a base layer for abrasive composite and strengthens the metallurgic bonding with the tool body and evens the distribution of CBN grains during the multi-layer sintering.

Laser beam influence on properties of CBN grain was investigated and optimal technological parameters of laser irradiation of CBN grains were established (power density $W_p=(0.1...1.8)x10^4$ W/cm^2, grain irradiation time - 0,2-0,3 sec depending on the size of sintered layer). With these parameters, the temperature interval of heating is kept within a range of 800-1400°C, and there is no negative thermal influence on CBN grains. It is recommended to use protective gases (argon, nitrogen) during sintering process at a flowrate of 5-14 l/min. It is also advisable to use grains with protective coating thus increasing thermal resistance and maintaining the abrasive properties of a separate grain.

There was developed a highly-efficient process of laser thermo-deformational sintering of CBN-contained abrasive composites, its technological principles were investigated and sample abrasive tools were manufactured. The discussed laser sintering process increases the productivity of abrasive tools manufacturing from 0,5-1 mm^2/min to 50-100 mm^2/min (ten times faster) and allows to use bonding material with various hardness that is varied from 300 to 10000 MPa.

ACKNOWLEDGMENTS

The article contains the results obtained within the framework of the grant support of NFDU under the competition "Science for Strengthening the Defense Capability of Ukraine" in 2024 (project registration number No. 2023.04/0051).

REFERENCES

Bing Guo, Qingyu Meng, Shuai Li, Guicheng Wu, Yang Xiang, Qingliang Zhao. 2022. Pulse laser precision truing of the V-shaped coarse-grained electroplating CBN grinding wheel, Materials & p. 110650.

Goncharuk, O., Zhuk, R., Kaglyak, O. 2018/ Laser Sintering of Abrasive Layers with Inclusions of Cubic Boron Nitride Grains. Lasers Manuf. Mater. Process. Vol. 5: 298–316.

Goncharuk, A. A.2017. The influence of laser radiation on the stability of cubic boron nitride. X International Scientific and Technical Web-Conference "Composite Materials"; p. 41–44.

Welding and Related Technologies – Krivtsun et al. (Eds)

Femtosecond laser infusion of graphite coating into aluminum foil

O. Kuznietsov & I. Gnilitskyi
Lviv Polytechnic National University, Lviv, Ukraine
"NoviNano Lab" LLC, Lviv, Ukraine

F. Ivashchyshyn
Lviv Polytechnic National University, Lviv, Ukraine
Czestochowa University of Technology, Czestochowa, Poland

ABSTRACT: This research investigates, for the first time, the possibility of fusing a thin graphite coating into aluminum foil using a femtosecond laser without altering its mechanical integrity. It is shown that by tuning the laser parameters, it is possible to achieve high-quality fusion of graphite to the surface. This is due to the unique interaction between the femtosecond laser pulses and the aluminum surface, which provides a high concentration of energy at local area in a short period of time. Further studies of the physicochemical processes during graphite infusion are important to optimize processing parameters such as laser modes and graphite performances. Extending this technology to other materials could pave the way to new opportunities in industry that require increased electrical conductivity and corrosion resistance. The results of this study can significantly influence the development of energy-saving technologies and the creation of high-performance materials for the electronics and energy storage fields.

Keywords: femtosecond laser processing, graphite coating, aluminum foil, supercapacitors, oxidation resistance, composite materials, laser-induced periodic surface structures (LIPSS)

1 INTRODUCTION

The research field of creating composites with carbon infusion into aluminum and other materials attracted significant attention from researchers over the world (Mohan 2018). The increasing demand for energy-efficient devices drives the search for new materials and technologies, particularly in developing high-efficient supercapacitors, crucial for energy storage and grid stability. Collectors in these devices require materials with high electrical conductivity and chemical resistance. Graphite-sprayed aluminum foil is promising due to its availability and conductive properties, but the poor adhesion between the graphite coating and the aluminum foil means that the coating is easily abraded and washed away. There are many various methods proposed to solve the problem, such as spark plasma sintering (Cavaliere 2017), low-temperature infiltration (Sasaki 2010), chemical vapor deposition (Doberdò 2014), hot pressing (Moghadam 2015), hot extrusion (Kwon 2012), mechanical mixing (Cavaliere 2017, Kwon 2012), and powder metallurgy (Kwon 2012, Moghadam 2015), combining extrusion and spark plasma sintering methods (Kwon 2009) to create such composites. Mentioned above studies have shown that carbon infusion into the aluminum matrix can significantly enhance the material's strength (Cavaliere 2017, Kimura 2020, Kwon 2012, Moghadam 2015, Sasaki 2010), electrical conductivity (Cavaliere 2017, Kwon 2009, Kwon 2012, Moghadam 2015), chemical resistance (Doberdò 2014, Kwon 2012), thermal conductivity (Cavaliere 2017), and wear resistance (Moghadam 2015). These improvements are explained by the formation of strong chemical bonds between carbon and aluminum, as well as the uniform distribution of

DOI: 10.1201/9781003518518-20

carbon in the matrix, which promotes better load transfer and corrosion resistance. It should be noted that laser infusion is much less frequently used for this purpose in comparison to other methods, but it still can be one of the promising strategies for infusing graphite coating into the surface (Kimura 2020).

Femtosecond laser treatment is a recent surface modification technique that enables precise nano- or micrometer structures with minimal contamination influence. Ultrafast lasers treatment have been explored for biomedical implants, sensors, optical holograms, and more, creating laser-induced periodic surface structures (LIPSS) with minimal thermal damage (Gnilitskyi 2021, Kuznietsov 2021).

Laser ablation on aluminum collectors increased the surface area and improved the mechanical adhesion of the electrode layers, reducing the interface resistance (Yang 2019). However, long-term tests showed capacity decreases due to aluminum oxide formation, highlighting the need for additional protection to ensure stability.

This study aims to determine the possibility of uniformly infusing graphite into the surface of aluminum foil using a femtosecond laser - a method not previously applied in this context to the best of our knowledge. Analysis using scanning electron microscopy (SEM) and energy-dispersive spectroscopy (EDS) examines the structure and graphite distribution in the material.

Although conductivity measurements were not conducted, this research explores the feasibility of laser-infused graphite as a potential future improvement. Successful infusion without damaging the base material could lead to effective collectors for supercapacitors with enhanced performance, applicable in energy storage systems, grid stabilization, and portable devices requiring rapid charge and discharge. These innovations could contribute to reducing energy costs and increasing efficiency, which is critically important for modern technological development.

2 EXPERIMENTAL TECHNIQUE

For the research, commercially available double-sided conductive aluminum foil (99.9% purity, 16 μm thickness) with a carbon coating (1 μm thickness and 0.5 g/m^2 density) used as a battery cathode substrate (MTI Corporation, California, USA).

The samples were treated with a Yb:KGW femtosecond laser (Pharos P-20, LightConversion, Vilnius, Lithuania). The schematic of the laser setup is shown in Figure 1 and was described in details in a previous research (Kuznietsov 2021).

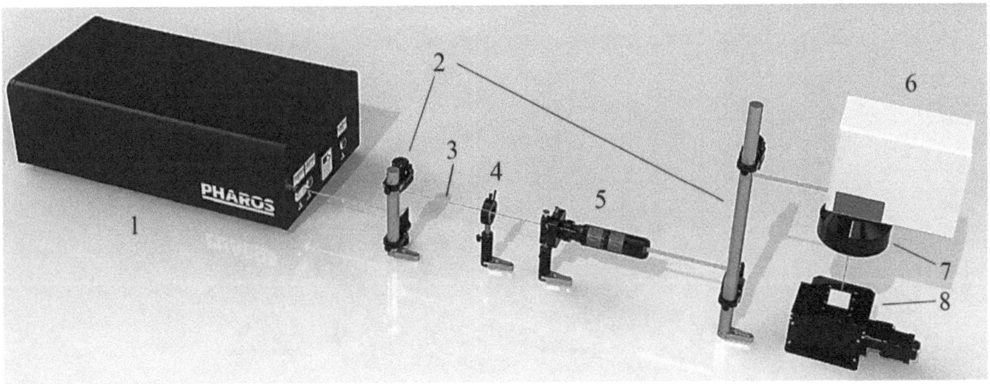

Figure 1. Experimental laser setup used in this work: 1 - femtosecond laser "Pharos", 2 - periscopes with mirrors, 3 - laser beam, 4 - half-wave plate, 5 - beam expander, 6 - galvanometer scanner, 7 - F-theta lens, 8 - motorized stage along the z-axis.

The processing was performed in the second harmonic mode at a wavelength of 515 nm, with a pulse duration of 266 fs and a pulse frequency of 400 kHz. The laser spot diameter was approximately 8 μm at the focus. The samples were processed in several modes (Table 1), with a scanning line spacing of 3 μm.

To remove the remains of unfused graphite from the surfaces of the samples, the remains of the graphite coating were carefully wiped with a cotton swab moistened with distilled water.

Table 1. Laser processing parameters for aluminum foil with graphite coating.

Sample	Average power, W	Energy, μJ	RR, kHz	Step, μm	Speed, m/s	Beam diametr, μm	Pulses per point	Fluens, J/sm^2
1	0,5	1,25			0,025		128	2,49
2	0,5	1,25	400	3	0,002	8	1600	2.49
3	0,64	1,6			0,1		32	3,19

The experiment results were studied using a scanning electron microscope (SEM) Phenom ProX (ThermoFisher Scientific, Massachusetts, USA). The elemental composition of the studied carbon was determined using an energy-dispersive spectroscopy (EDS) module in this microscope.

Surface roughness measurements were made using a non-contact profilometer Sensofar S lynx (Sensofar, Terrassa, Spain). Measurements were carried out by the confocal method at 150-fold magnification with a height step of 0.1 μm. The analysis of the surface topography was carried out according to the ISO 25178 standard using areal Gaussian filters.

3 RESULTS AND DISCUSSION

Using EDS and roughness (Sa) measurements, five random points were analyzed for each of the graphite processing modes fused into aluminum foil using femtosecond laser processing (Figure 2). In Mode #1 (Figure 2a), uniform carbon deposition with some oxidation was achieved. SEM and EDS analyses showed high-quality LIPSS and islands of fused silica formed on the surface, increasing the roughness from 0.048867 μm (untreated foil) to 0.2027 μm, compared to the roughness of sputtered untreated graphite at 0.2084 μm. Calcium contamination was detected at one of the EDS measurement points, likely because the standard laboratory conditions for laser processing do not meet cleanroom requirements.

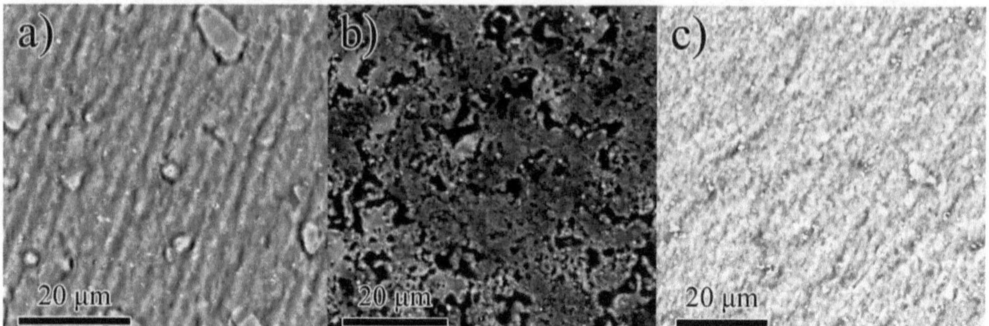

Figure 2. SEM image of the results of femtosecond laser processing. a) The result of processing the sample #1; b) sample #2 and c) sample #3.

In Mode #2 (Figure 2b), increasing the number of pulses per point led to a higher proportion of filled graphite and the highest surface roughness (0.2965 μm), even exceeding the roughness of the sputtered untreated graphite powder. Significant thermal effects caused the greatest surface oxidation compared to other modes, which may negatively affect the material's conductivity despite the increased graphite content and high roughness.

In Mode #3 (Figure 2c), increasing the power and decreasing the number of pulses per point reduced the amount of oxide formed on the surface, but at the same time, a smaller proportion of carbon is observed than in other modes. Additionally, the sample processed in this mode has the lowest roughness among the other samples (0.099859 μm). This may be due to stronger material ablation during laser processing, caused by the higher laser radiation power.

These results demonstrate that by adjusting the laser processing parameters, one can significantly influence the surface roughness, oxidation levels, and graphite content in aluminum foil, depending on the requirements.

4 CONCLUSIONS

Femtosecond laser fusion of graphite to aluminum foil successfully forms uniform carbon layers and laser-induced periodic surface structures (LIPSS), increasing surface area and adhesion with precise control and minimal thermal effects. However, oxidation during processing in air can have the negative effect of increasing surface resistance. Vacuum or inert gas treatment may be suggested to ensure long-term stability and prevent oxidation.

REFERENCES

Cavaliere, P., Sadeghi, B., & Shabani, A. 2017. Carbon nanotube reinforced aluminum matrix composites produced by spark plasma sintering. *Journal of Materials Science* 52: 8618–8629.

Doberdò, I., Löffler, N., Laszczynski, N., Cericola, D., Penazzi, N., Bodoardo, S. & Passerini, S. 2014. Enabling aqueous binders for lithium battery cathodes–Carbon coating of aluminum current collector. *Journal of power sources* 248: 1000–1006.

Gnilitskyi, I., Mamykin, S. V., Lanara, C., Hevko, I., Dusheyko, M., Bellucci, S., & Stratakis, E. 2021. Laser nanostructuring for diffraction grating based surface plasmon-resonance sensors. *Nanomaterials* 11(3): 591.

Kimura, T., Nakamoto, T., Suyama, T., & Miki, T. 2020. In-process fabrication of carbon-dispersed aluminum matrix composite using selective laser melting. *Metals* 10(5): 619.

Kuznietsov, O. V., Tsibidis, G. D., Demchishin, A. V., Demchishin, A. A., Babizhetskyy, V., Saldan, I., Bellucci, S. & Gnilitskyi, I. 2021. Femtosecond laser-induced periodic surface structures on 2D Ti-Fe multilayer condensates. *Nanomaterials* 11(2): 316.

Kwon, H., Estili, M., Takagi, K., Miyazaki, T., & Kawasaki, A. 2009. Combination of hot extrusion and spark plasma sintering for producing carbon nanotube reinforced aluminum matrix composites. *Carbon* 47(3): 570–577.

Kwon, H., & Leparoux, M. 2012. Hot extruded carbon nanotube reinforced aluminum matrix composite materials. *Nanotechnology* 23(41): 415701.

Moghadam, A. D., Omrani, E., Menezes, P. L., & Rohatgi, P. K. 2015. Mechanical and tribological properties of self-lubricating metal matrix nanocomposites reinforced by carbon nanotubes (CNTs) and graphene–a review. *Composites Part B: Engineering* 77: 402–420.

Mohan, V. B., Lau, K. T., Hui, D., & Bhattacharyya, D. 2018. Graphene-based materials and their composites: A review on production, applications and product limitations. *Composites Part B: Engineering* 142: 200–220.

Sasaki, G., Hara, Y., Xu, Z. F., Sugio, K., Fukushima, H., Choi, Y. B., & Matsugi, K. 2010. Fabrication of carbon nano-fiber/aluminum composites by low-pressure infiltration method. *Materials Science Forum* 654: 2692–2695.

Yang, D., & Laforgue, A. 2019. Laser surface roughening of aluminum foils for supercapacitor current collectors. *Journal of the Electrochemical Society* 166(12): A2503.

Welding and Related Technologies – Krivtsun et al. (Eds)
© 2025 The Author(s), ISBN 978-1-032-85176-1
Open Access: www.taylorfrancis.com, CC BY-NC-ND 4.0 license

Workplace of a cosmonaut-welder for performing manual electron beam welding of objects in open space

S.O. Hlushak, O.R. Bulatsev, N.V. Piskun, Ye.H. Ternovy & V.A. Kryukov
E.O. Paton Electric Welding Institute of the NAS of Ukraine, Kyiv, Ukraine

Zeng Ruchuan
Beijing Spacecrafts, Beijing, P.R.C., China

ABSTRACT: Stabilizing the body of cosmonaut-welder enables precisely performing the movements which ensure sound welding in orbit. For this purpose it is necessary to create a device, which will allow rigid fixation in the required position before performing the technological operations. A workplace of welding operator was developed, which is designed for performance of technological and repair work during operation of the station modules in open space or on the Moon surface. It is also necessary for movement of the operator to the location of performance of operations on maintenance and repair of the space module sheath with the possibility of its lifting, tilting and rotation around a vertical axis.

1 INTRODUCTION

When mastering the near-Earth space and Moon surface, it will be necessary to perform mounting and, in a number of cases, repair-restoration work. Here, welding is a priority technological process during fabrication of structures for space applications, which allows producing reliable joints of structural elements with guaranteed sealing without lowering their strength and with preservation of their weight and size characteristics.

It is known that the hardware and instrumentation planned to be used or tested in space should pass numerous preflight tests in a flying laboratory under microgravity conditions and in a thermal pressure chamber at temperatures of ± 120°C.

At operation in the thermal pressure chamber, i.e. under vacuum, for a person dressed in a spacesuit, the coordination of the arms becomes extremely complicated, the flow of information is reduced, which, in turn, dramatically affects the result of the work of the welding operator.

Therefore, one of the main problems is stabilizing the body of a cosmonaut-welder in order to accurately perform sufficiently coordinated and complex movements that ensure high-quality welding in orbit. More over, alongside development of specialized electron beam welding hardware to perform technological operations in open space, it is also necessary to create auxiliary equipment, which is the welding operator workplace (WP).

The main technical characteristics of welding operator workplace are as follows: shipping dimensions, weight characteristics, service area, selection of the design of mobile workplace chassis, as well as specific selection of the electric drive and its power supply (autonomous or from a solar cell, or from the space station network).

The E.O.Paton Electric Welding Institute has many years of experience of development, creation and testing in space of several modifications of welding operator workplace, which are described below.

DOI: 10.1201/9781003518518-21

2 DEVELOPMENT OF WELDING OPERATOR WORKPLACE

2.1 *Need to create the workplace*

When preparing the world's first experiment on welding and other technologies in open space the welding researchers faced the problem which required an immediate solution. When retro-fitting the hardware, it was necessary to conduct technological experiments on welding. There was very little time for performance of these experiments. In order to create a state of weightlessness, the plane which was equipped with the necessary welding hardware, flew along the Kepler curve. The state of weightlessness lasted for just several seconds. During this time, the welder in a spacesuit was to perform complex and very precise welding work.

The first experiments on welding, which we tried to carry out under the conditions close to the space ones showed that the welder- very often looses the welding tool, particularly under microgravity. He is not able to stay fixed in place and to perform coordinated movements during welding, while holding the dangerous electron beam gun.

It was quite obvious that the problem of stabilizing the body of cosmonaut-welder, possibility of precise performance of quite complex movements by him, which ensure high-quality welding in orbit, are some of the most complicated tasks in the overall complex for welding in space.

During analysis of flight tests the welding researchers together with Boris. E. Paton came to the conclusion that it is necessary to develop a device, which will allow rigidly fixing the body in one position and without moving, concentrate on performance of the technological process.

2.2 *Performance of the first experiments under the conditions of open space*

During development of welding operator workplace, in addition to prevention of cosmonaut movement, not rigidly attached to the vehicle hull, it was also necessary to take into account that during performance of different operations in space, it is necessary to apply tools with holders, absorbing the torque, or special tools with very small torque, acting on the cosmonaut. And where possible, it is rational to perform complex operations at rigid fixation of the operator to the vehicle hull. Mechanical folding or sliding devices, magnetic and adhesive substances can be used for this purpose.

The first workplace of cosmonaut-welder was developed for performance of experiments in open space on board the Salyut-7 station. It was made according to all safety and ergonomic requirements.

Figure 1 shows how the cosmonaut performs coating deposition in open space outside Salut-7 orbital station. (Hlushak 2023).

Figure 1. Cosmonaut-welder performs the technological operations of welding, cutting and coating deposition for the first time in open space on board of Salyut-7 orbital station. (Glushak 2023).

The cosmonaut-welder workplace allowed performing the technological operations of welding, brazing, cutting and coating deposition for 3 hours and 35 minutes in open space. WP was oriented in such a way relative to the side of Salyut-7 station that when the electron beam unauthorizedly cuts a hole in the sample, or it will pass by the board with the samples, the beam should automatically switch-off so as not to burn through the space station hull.

This experiment showed, first, the possibility of performance of the operations of welding and related technological processes with the electron beam tool in open space, secondly, that the operator wearing a space suit, can safely perform complicated manual technological operations, when using the workplace, and obtain sound results.

3 MPESS WORK PLARFORM FOR CONDUCTING THE INTERNATIONAL SPACE WELDING EXPERIMENT (ISWE)

Generalization of the accumulated experience allowed development of an all-purpose complex of electron beam tool «Universal», designed for operation as part of future large-sized long-term orbital space stations.

During preparation of the international welding experiment the E.O.Paton Institute designed and manufactured the hardware which is a special platform (Figure 2a) (Paton et al. 2000). Attached to the platform was the hardware required for experiment performance (drum carrying the samples and container for replaceable samples), and the adjusted operator workplace (Figure 2b). All that - the cosmonaut-welded workplace was developed to accommodate the US spacesuit.

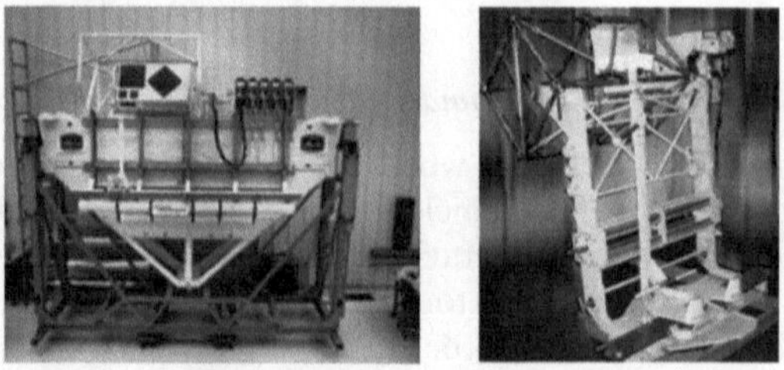

a б

Figure 2. General view of MPESS platform to conduct ISWE.
(a) Mounting ISWE hardware mock-up in the berth on Columbia payload platform (Paton et al. 2000)
(b) Cosmonaut workplace with MPESS platform.

In 1991–1992 Universal hardware was used to perform a series of technology trials on welding samples from aluminium and titanium alloys, as well as stainless steels on the ground, as well as under microgravity conditions in KC-135 flying laboratory (L.Johnson Center, NASA, Huston), as well as six immersions in a water immersion facility (at G.Marshall Center, NASA, Huntsville) and five ascents to 6 km in a pressure chamber.

Appropriate changes were made in the design of this auxiliary equipment to enhance the reliability during experiment performance and to strictly observe the safety regulations.

Figure 3 shows preflight tests of the hardware using the welding operator mobile workplace.

In 1998 it was planned to conduct in the "Mir" station the welding experiment, which was not performed in Columbia spacecraft. A new workplace was designed and manufactured for this experiment named "Flagman". In view of the limited work space at "Mir" station, the workplace was very compact, but fitted with all the necessary elements for experiment performance, and it was called the "Work station".

Figure 4 shows two models of the work station: first one, later on improved, which was shipped to "Mir" station (Paton et al. 2000).

Two patents for invention were granted for the welding operator workplace in the USA and Ukraine (Figure 5).

Availability of patents once more confirms the importance of development and application of this equipment.

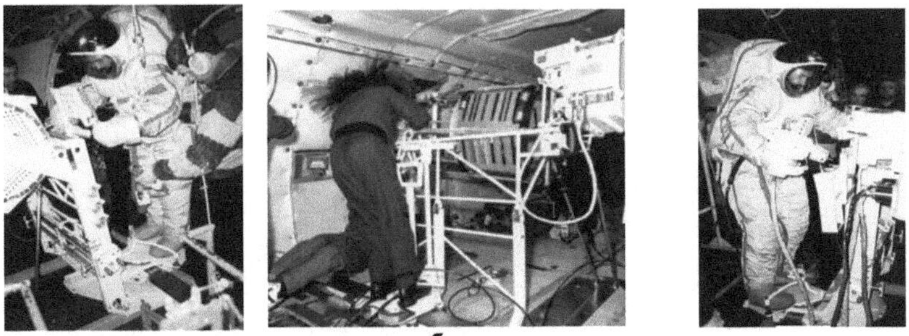

a б в

Figure 3. Preflight tests of the hardware using the workplace.

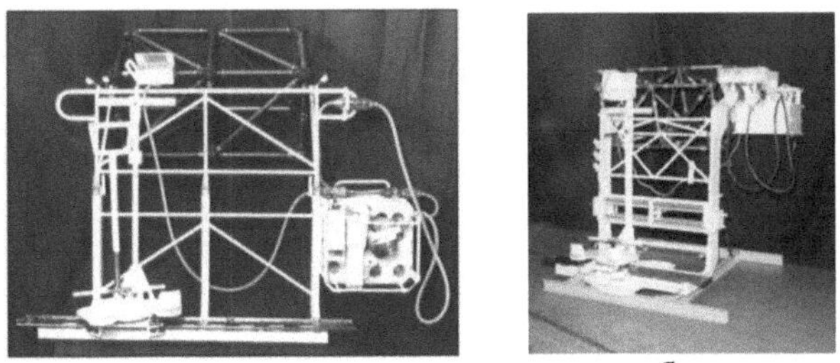

a б

Figure 4. Work station for performance of welding experiment on «Mir» station: a- first model of the work platform (Paton et al. 2000); b- work station (Glushak 2023).

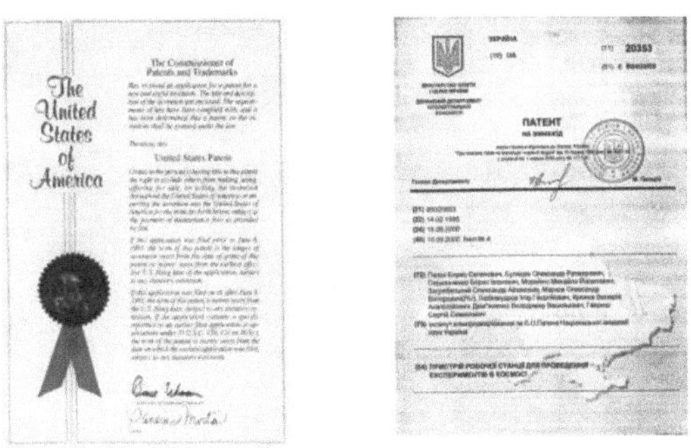

Figure 5. Patents for cosmonaut workplace.

4 CONCLUSIONS

During performance of operations in open space and on the Moon surface, in order to conduct the process of electron beam welding, the specialized WP of cosmonaut-welder must be part of the overall hardware complex for working in space. Solving this problem allows fixing the body of cosmonaut-welder that is required for precise coordination of complex movements, ensuring sound and stable manual arc welding in orbit or on the Moon surface.

REFERENCES

S.O. Hlushak (2023) Evolution of electron beam hardware for welding in space. The Paton Welding J., 8, 78–84. DOI: https://doi.org/10.37434/tpwj2023.08.11.

Paton et al. 2000. Work on Manual Electron Beam Technology in Space. *Space: Technology, Materials Science, Constructions. Collection of scientific papers under. edited by Academician B.E. Paton.* Kyiv: PWI them. E. O. Paton NAS of Ukraine, 190–203.

Session 4. Mathematical modeling of welding and related processes

Welding and Related Technologies – Krivtsun et al. (Eds)

Modelling of tungsten inert gas arc and its impact on the anode

I.V. Krivtsun, V. Demchenko, I. Krikent & O. Semenov
The E.O. Paton Electric Welding Institute of the NASU, Kyiv, Ukraine

Shiyi Gao & Boyi Wu
China-Ukraine Institute of Welding, Guangdong Academy of Sciences, Guangdong Provincial Key Laboratory of Advanced Welding Technology, Guangzhou, China

ABSTRACT: An unified model of plasma column and anode region of the electric arc with refractory cathode burning in inert gas is described. This model allows to take into account the non-uniform distribution of anode voltage fall in the region of anode binding, which makes description of the near-anode plasma more adequate. The model can be used for calculation of thermal, gas-dynamic and electromagnetic characteristics of arc plasma and characteristics of its thermal, electric and force impact on the anode surface, depending on arc current and length, kind of inert gas and anode material. Testing and verification of the model has been carried out, the results of which indicate its applicability to the numerical study of the above characteristics.

1 INTRODUCTION

Efficient application of the electric arc as a welding heat source requires valid information about its thermal, electrical and force effects on the weld metal surface. In DC TIG welding such effects are mainly determined by the physical processes in plasma column and anode region of the arc and depend on its current and length, shape of the cathode tip, shielding gas composition and pressure (Nestor 1962). Since experimental study of welding arc is difficult, this paper is aimed to mathematical modelling of the TIG arc, scheme of which is presented in Figure 1.

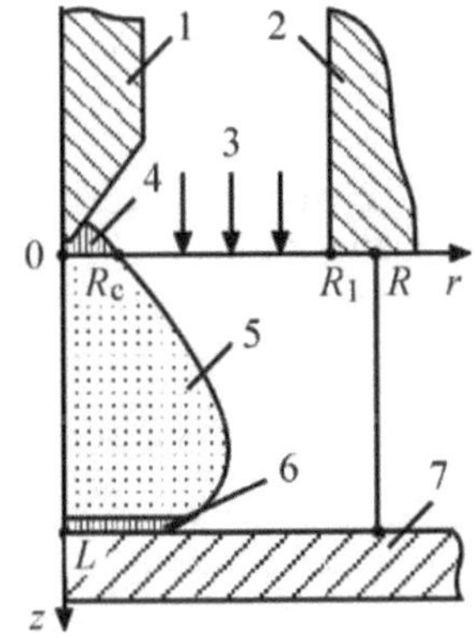

Figure 1. Schematic for modelling: 1 – tungsten cathode; 2 – nozzle; 3 – shielding inert gas; 4 – cathode region; 5 – arc column; 6 – anode boundary layer; 7 – anode; R_c – cathode region radius; R_1 – nozzle radius; R – radius of calculation domain; L – arc length.

DOI: 10.1201/9781003518518-22

In the majority of publications on complex modelling of TIG arcs (Hsu et al. 1983, Goodarzi et al.1997, Mougenot et al. 2013) the simplified models of the near-electrode regions are usually used, while the studies devoted to the anode and cathode phenomena (Dinulescu & Pfender 1980, Almeida et al. 2008, Semenov et al. 2016) do not pay sufficient attention to their interrelation with the processes in the arc column.

Since the theory of cathode phenomena in TIG arcs has been developed in detail (Almeida et al. 2008), we will focus on the processes in the arc column and anode region, using the recommendations (Wendelstorf et al. 1997) as boundary conditions near the cathode and will use the approach proposed in (Krivtsun et al. 2015). Accordingly, the unified mathematical model of the plasma system considered should include two interconnected models: 1. Model of arc column, describing processes in the arc column plasma; 2. Model of anode boundary layer, allowing formulation of boundary conditions on the interface between arc column and anode region.

2 MODEL OF ARC COLUMN

The following assumptions have been made in formulating equations of this model:

- arc plasma is in the stationary state and is assumed to be axisymmetric;
- arc column plasma flow is viscous and subsonic, flow mode is laminar;
- gravitational force is neglected, external magnetic field is absent;
- the main mechanism of arc column plasma heating is Joule heating;
- energy transfer is performed due to thermal conductivity, convection and electron drift;
- arc plasma is optically thin for plasma radiation.

The governing equations for mass, momentum, energy and charge transfer in the arc column plasma, written in the cylindrical coordinates (r, ϑ, z) are as follows:

$$\frac{1}{r}\frac{\partial}{\partial r}(r\rho v) + \frac{\partial}{\partial z}(\rho u) = 0, \tag{1}$$

where ρ = arc plasma density; v, u = radial and axial components of plasma velocity.

$$\rho\left(v\frac{\partial v}{\partial r} + u\frac{\partial v}{\partial z}\right) = \frac{-\partial p}{\partial r} - j_z B_\vartheta + \frac{2}{r}\frac{\partial}{\partial r}\left(r\eta\frac{\partial v}{\partial r}\right) + \frac{\partial}{\partial z}\left[\eta\left(\frac{\partial u}{\partial r} + \frac{\partial v}{\partial z}\right)\right] - 2\eta\frac{v}{r^2} - \frac{2}{3}\frac{\partial}{\partial r}\left\{\eta\left[\frac{1}{r}\frac{\partial(rv)}{\partial r} + \frac{\partial u}{\partial z}\right]\right\};$$

$$\rho\left(v\frac{\partial u}{\partial r} + u\frac{\partial u}{\partial z}\right) = \frac{-\partial p}{\partial z} + j_r B_\vartheta + 2\frac{\partial}{\partial z}\left(\eta\frac{\partial u}{\partial z}\right) + \frac{1}{r}\frac{\partial}{\partial r}\left[r\eta\left(\frac{\partial u}{\partial r} + \frac{\partial v}{\partial z}\right)\right] - \frac{2}{3}\frac{\partial}{\partial z}\left\{\eta\left[\frac{1}{r}\frac{\partial(rv)}{\partial r} + \frac{\partial u}{\partial z}\right]\right\}, \tag{2}$$

where p = pressure; j_r, j_z = radial and axial components of the electric current density; B_ϑ = azimuthal component of the magnetic field; η = dynamic viscosity of arc plasma.

$$\rho C_p\left(v\frac{\partial T}{\partial r} + u\frac{\partial T}{\partial z}\right) = \frac{1}{r}\frac{\partial}{\partial r}\left(r\chi\frac{\partial T}{\partial r}\right) + \frac{\partial}{\partial z}\left(\chi\frac{\partial T}{\partial z}\right) + \frac{k}{e}\left[j_r\frac{\partial(\alpha T)}{\partial r} + j_z\frac{\partial(\alpha T)}{\partial z}\right] + \frac{j_r^2 + j_z^2}{\sigma} - \psi. \tag{3}$$

Here C_p = heat capacity of arc plasma; T = plasma temperature; χ, σ = thermal and electrical conductivities respectively; e, k = elementary charge and the Boltzmann constant; α = $5/2 - \delta$, where δ = thermal diffusion constant of electrons; ψ = radiation energy losses.

$$\frac{1}{r}\frac{\partial}{\partial r}\left(r\sigma\frac{\partial\varphi}{\partial r}\right) + \frac{\partial}{\partial z}\left(\sigma\frac{\partial\varphi}{\partial z}\right) = 0;\ B_\vartheta(r,z) = \frac{\mu^0}{r}\int_0^r j_z(r',z)r'dr';\ j_r = -\sigma\frac{\partial\varphi}{\partial r};\ j_z = -\sigma\frac{\partial\varphi}{\partial z}, \tag{4}$$

where φ = electric potential, μ^0 = universal magnetic constant.

Assuming that anode material evaporation is neglected and arc column plasma is in local thermodynamic equilibrium, we will use the following equations to calculate of its composition:

$$\frac{n_e n_1}{n_0} = \left(\frac{2\pi m_e kT}{h^2}\right)^{3/2} \frac{2\theta_1}{\theta_0} exp\left(\frac{-eU_i}{kT}\right); \; n_e = n_1; \; p = (n_e + n_0 + n_1)kT. \tag{5}$$

Here n_e, n_Z = number densities of electrons, atoms (Z = 0) and ions (Z=1) of shielding inert gas; θ_Z = their statistical sums; m_e, h = electron mass and the Planck constant; U_i = ionization energy.

3 MODEL OF ANODE BOUNDARY LAYER

We will assume that non-equilibrium plasma of the anode boundary layer is characterized by electron temperature T_e and heavy particle temperature $T_h < T_e$, which are assumed to be constant across this layer, but dependent on r:

$$T_e(r) = T(r, L); \; T_h(r) = T_s(r), \tag{6}$$

where $T(r, L)$ = distribution of the arc column plasma temperature on the interface with anode boundary layer; $T_s(r)$ = distribution of the anode surface temperature.

It is known that the anode potential fall $U_a = \varphi_a - \varphi_{pa}$ is negative for atmospheric pressure arcs (Sanders & Pfender 1984, Tanaka & Ushio 1999) and, according to (Krivtsun et al. 2015), has non-uniform distribution along the anode surface. Here φ_a = potential of the anode, which is assumed to be constant in view of high conductivity of anode metal, and is selected to be zero; $\varphi_{pa}(r)$ = potential of the arc plasma at the interface with the anode boundary layer, to calculate of which the following expression can be used:

$$\varphi_{pa}(r) = \frac{kT_e(r)}{e} ln\left\{\frac{en_e(r)\bar{v}_e(r)}{4[j_a(r) + j_i(r)]}\right\}. \tag{7}$$

Here $n_e(r) = n_e(r, L)$, $\bar{v}_e(r) = [8kT_e(r)/\pi m_e]^{1/2}$ are the distributions of number density and average speed of chaotic (thermal) motion of plasma electrons on this interface; $j_a(r) = |j_z(r, L)|$ is the absolute value of the electric current density on the anode; $j_i(r) = en_e(r)exp(-1/2)V_B(r)$ is the ion current density from arc plasma to the anode surface, where $V_B(r) = \{k[T_e(r) + T_h(r)]/M_i\}^{1/2}$ is the local value of the Bohm velocity; M_i = ion mass.

We will present the heat flux from arc plasma to the anode surface as $q_{pa} = q_e + q_i$, where q_e, q_i = fluxes of energy, transferred by plasma electrons and ions respectively. We will use the following expressions to calculate their radial distributions:

$$q_e(r) = \frac{5}{8} n_e(r)\bar{v}_e(r)kT_e(r)exp\left[\frac{-e\varphi_{pa}(r)}{kT_e(r)}\right];$$

$$q_i(r) = j_i(r)\left[\varphi_{pa}(r) - \frac{1}{2}\frac{kT_e(r)}{e} + \frac{M_i V_B^2(r)}{2e} + \beta_r U_i\right]. \tag{8}$$

Here β_r is the coefficient of ion recombination energy accommodation by the anode surface. It should be noted that the distribution of the total heat flux, contributed by the arc to the anode, should be calculated allowing for work function of electrons ς_a of the anode metal, that yields

$$q_a(r) = q_{pa}(r) + j_a(r)\varsigma_a. \tag{9}$$

Thus, the anode boundary layer model allows calculation of the distributions of anode voltage drop $U_a(r) = -\varphi_{pa}(r)$ and heat flux into the anode $q_a(r)$, depending on the distributions of $j_z(r, L)$ and $T(r, L)$, which can be determined from the model of arc column.

4 BOUNDARY CONDITIONS

Near the cathode tip (plane $z = 0$) the boundary conditions for the plasma velocity are as follows:

$$v|_{z=0} = 0; u|_{z=0} = \begin{cases} u_0, 0 \le r \le R_1; \\ 0, R_1 < r \le R, \end{cases} \tag{10}$$

where u_0 is determined by shielding gas feed rate. We will use the following conditions for the temperature and electrical potential in the cathode region ($0 \le r \le R_c$):

$$T|_{z=0} = T_c(r); \ \sigma\frac{\partial\varphi}{\partial z} = j_c(r), \tag{11}$$

where $T_c(r), j_c(r)$ = distributions of plasma temperature and arc current density near the cathode (Wendelstorf et al. 1997). Outside the cathode region ($R_c \le r \le R$) we set:

$$T|_{z=0} = T_0; \ \frac{\partial\varphi}{\partial z} = 0. \tag{12}$$

Here T_0 = temperature of the shielding gas inputted.

Since the thickness of the anode boundary layer in atmospheric pressure arc discharges is significantly less than the characteristic dimensions of the arc column (Dinulescu & Pfender 1980), this layer can be considered infinitely thin. The energy balance of the anode boundary layer can be used as the boundary condition for temperature on its interface with the arc column plasma, or, allowing for the above mentioned assumption, at $z = L$

$$\left[-\chi\frac{\partial T}{\partial z} - j_z\frac{k}{e}\alpha T\right]\Bigg|_{z=L} = \varphi_{pa}(r)j_a(r) + q_{pa}(r). \tag{13}$$

Expression in the left hand side of (15) describes power, supplied to the anode boundary layer from the arc column. The first term in the right hand side describes power, consumed for maintaining this layer, and the second one is heat flux towards the anode surface. The boundary condition for the electric potential on the interface of plasma column with anode boundary layer is

$$\varphi|_{z=L} = \varphi_{pa}(r). \tag{14}$$

For the velocity field, no slip boundary conditions are used on the anode surface:

$$v|_{z=L} = u|_{z=L} = 0. \tag{15}$$

The assumption of axial symmetry gives the next boundary conditions on the axis ($r = 0$):

$$v|_r = 0 = \frac{\partial u}{\partial r}\bigg|_{r=0} = \frac{\partial T}{\partial r}\bigg|_{r=R} = \frac{\partial \varphi}{\partial r}\bigg|_{r=0} = 0. \tag{16}$$

On the outer boundary ($r = R$) for the velocity and electric potential we can write:

$$u|_{r=R} = \frac{\partial(r\rho v)}{\partial r}\bigg|_{r=R} = \frac{\partial \varphi}{\partial r}\bigg|_{r=R} = 0 \tag{17}$$

and for the plasma temperature, depending on the direction of its radial velocity, we set:

$$T|_{r=R} = T_0,\ v|_{r=R} \le 0;\ \frac{\partial T}{\partial r}\bigg|_{r=R} = 0,\ v|_{r=R} > 0. \tag{18}$$

5 TESTING AND VERIFICATION OF THE MODEL

Numerical analysis of TIG arc with water-cooled steel/copper anode in argon of atmospheric pressure was carried out. The equations (1)-(4) with the boundary conditions (10)-(18) were solved numerically by finite difference method using the in-house code developed. Thermodynamic, transport and optical properties of argon plasma were taken from (Cressault et al. 2013, Essoltani et al. 1994). The simulation results were compared with the experimental data (Nestor 1962, Sanders & Pfender 1984) and the calculated ones (Semenov et al. 2013).

The calculated values of the electric potential φ_{pa} depending on the near-anode plasma temperature and anode current density for a steel anode are shown in Figure 2. The anode potential fall $U_a = -\varphi_{pa}$ is negative, its absolute value increases with rise of T_e (Figure 2a) and decreases as j_a grows (Figure 2b). At $j_a = 5\times10^6$ A/m^2 and $T_e = 8000$ K the model used gives $U_a = -1.96$ V, while the corresponding value, calculated by the model (Semenov et al. 2013), equals -2.13 V.

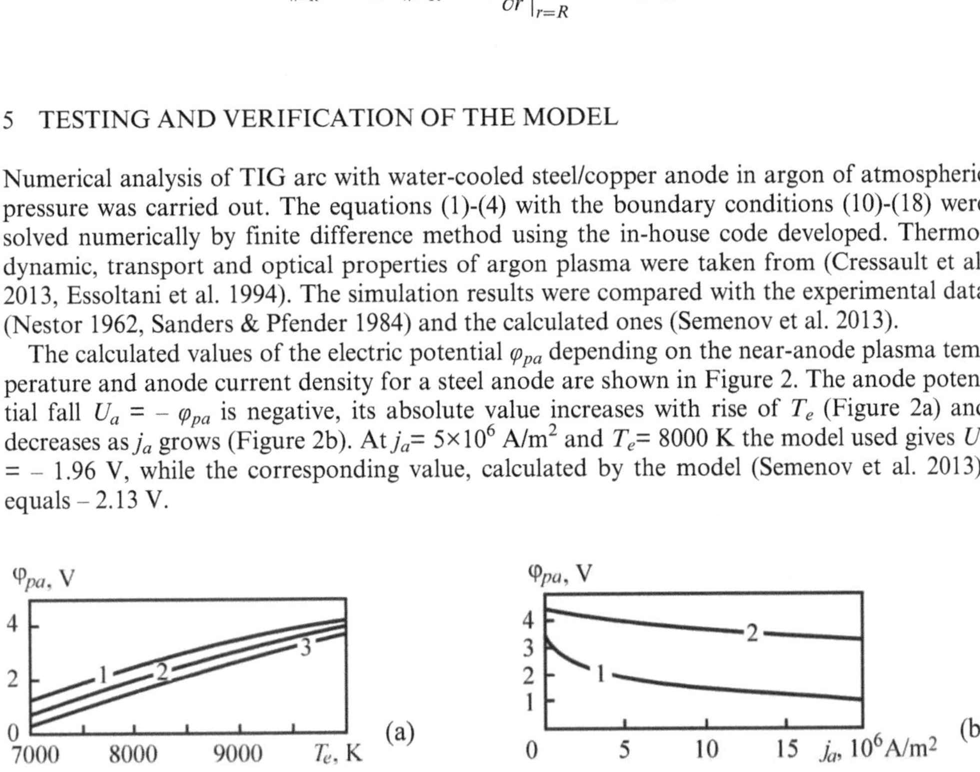

Figure 2. Electric potential of arc column plasma on the interface with anode boundary layer, depending on T_e (a) and j_a (b): a – 1 – $j_a = 2\times10^6$, 2 – 5×10^6, 3 – 10^7 A/m^2; b – 1 – $T_e = 8000$, 2 – 10000 K.

The dependencies of the heat flux q_a on T_e and j_a are incremental (Figure 3). At $j_a = 5\times10^6$ A/m^2 and $T_e = 8000$ K, our calculations gives $q_a = 39.5\times10^6$ W/m^2. The corresponding heat flux into the anode calculated in (Semenov et al. 2013) is 27.7×10^6 W/m^2, which is sufficiently smaller.

According to the probe measurements (Sanders & Pfender 1984) for argon arc with copper water-cooled anode ($I = 200$ A; $L = 10$ mm), electron temperature on the anode boundary layer axis $T_e(0) = 9840$ K. The experimental data (Nestor 1962) give $j_a(0) \approx 3.5\times10^6$ A/m^2. Using these data, we have found that $U_a(0) = -4.08$ V. This value corresponds with a high accuracy to $U_a = -4.01$ V, determined experimentally (Sanders & Pfender 1984).

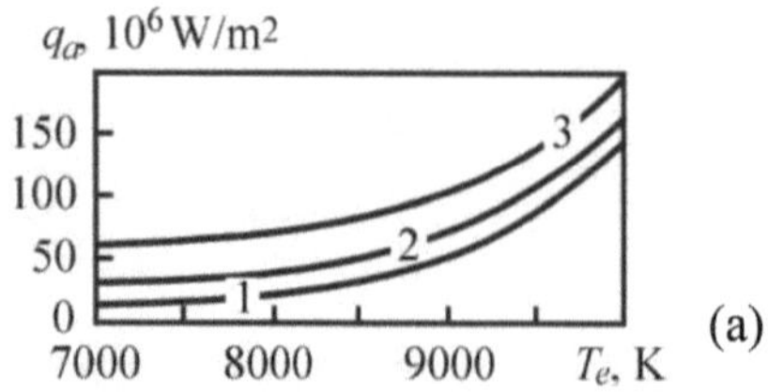

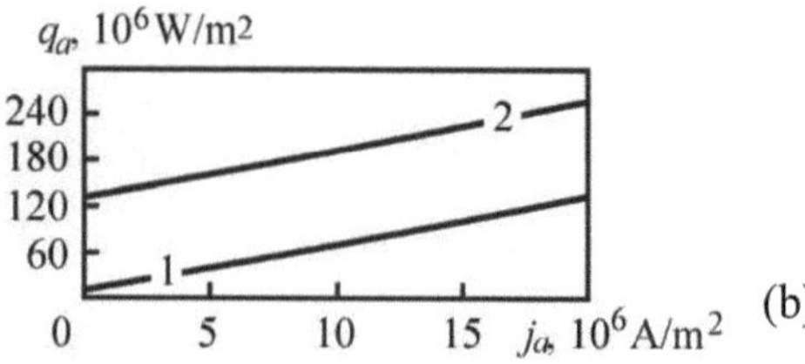

Figure 3. Dependences of heat flux into the anode on electron temperature (a) and current density (b) in the anode boundary layer: a – 1 – $j_a = 2\times10^6$, 2 – 5×10^6, 3 – 10^7 A/m^2; b – 1 – T_e = 8000, 2 – 10000 K.

Comparison of calculated radial distributions of j_a and q_a with the experimental results (Nestor 1962) for argon arcs with copper water-cooled anode (I = 100, 200 A; L = 6.3 mm) is shown in Figure 4. The calculated integral heat input into anode for 100 and 200 A arcs are 894 and 1982 W, the corresponding values obtained using experimental data (Nestor 1962) are 1007 and 1871 W. As can be seen, a sufficiently good agreement of calculated and experimental data is observed, especially at I = 200 A.

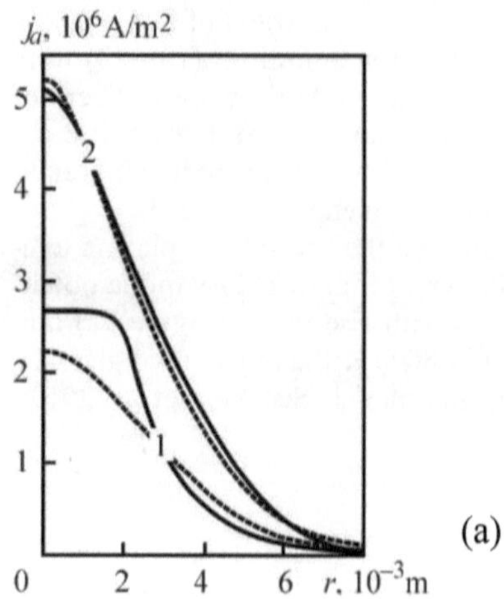

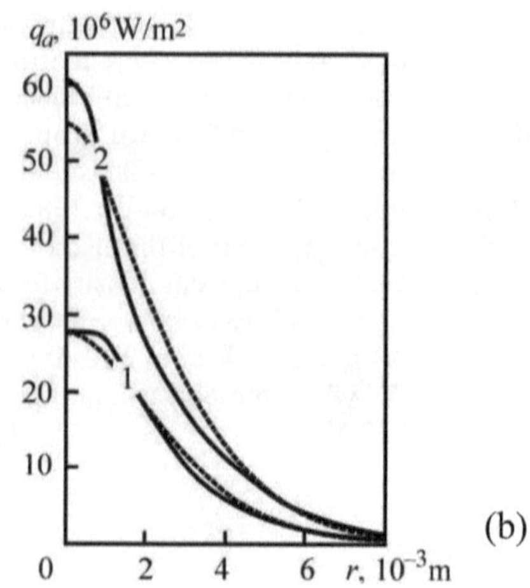

Figure 4. Radial distributions of anode current density (a) and heat flux into anode (b): 1 – I = 100; 2 – 200 A; dotted curves – calculation; solid curves – experimental data (Nestor 1962).

To analyze the effect of non-uniform distribution of φ_{pa} on characteristics of the near-anode plasma and its impact on the anode, two cases are considered: $\varphi_{pa}(r)$ is calculated with the help of (7); φ_{pa} is chosen to be constant and equals $\varphi_{pa}(0)$ obtained in the first case. In the first case $\varphi_{pa}(r)$ reaches its maximum 2.62 V on the arc axis (Figure 5a) due to the high temperature in this region (Figure 5b). The temperature distributions are practically don't differ in the first and second cases. At the same time the radial distributions of j_a and q_a in the second case differ substantially both from those obtained in the first one (Figure 6) and from the experimental data (Nestor 1962).

To explain this effect, let's consider the radial component of current density in the near-anode plasma (Figure 7). In the first case, at $z > L - 0.25$ mm $j_r > 0$ (Fig. 7a), i.e. is directed away from the arc axis. This means that the absolute value of $j_z(0)$ increases to the anode surface. In the second case, j_r is negative for all z (Figure 7b), resulting in a decreasing of $|j_z(0)|$ towards the anode. These facts result in the calculated values of $j_a(0)$ and $q_a(0)$ being significantly higher in the first case than in the second one (Figure 6). Thus, by taking into account the non-uniformity of U_a, it is possible to describe the effect of the arc current contraction in the near-anode plasma.

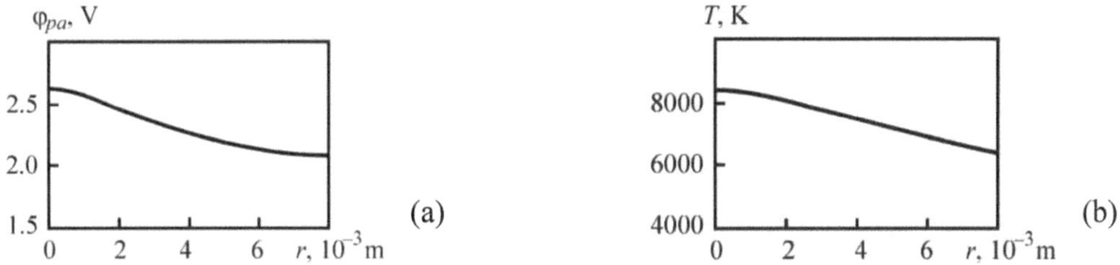

Figure 5. Radial distributions of arc plasma potential (a) and temperature (b) on the interface between anode boundary layer and arc column (I = 200 A; L=6.3 mm).

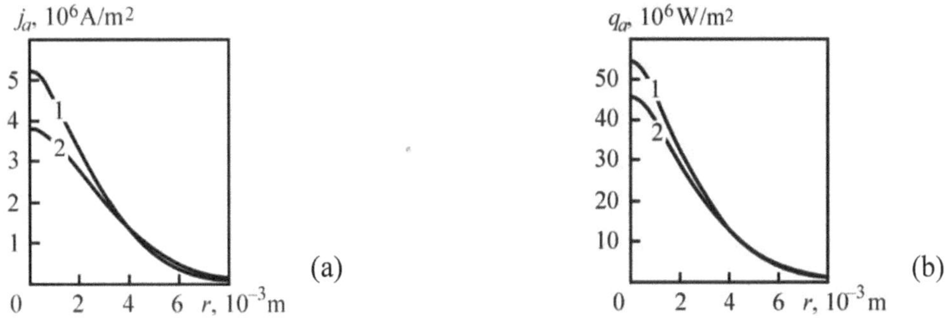

Figure 6. Radial distributions of anode current density (a) and heat flux (b): 1 – calculation, taking into account dependence $\varphi_{pa}(r)$; 2 – calculation at φ_{pa} = 2.62 V (I = 200 A; L=6.3 mm).

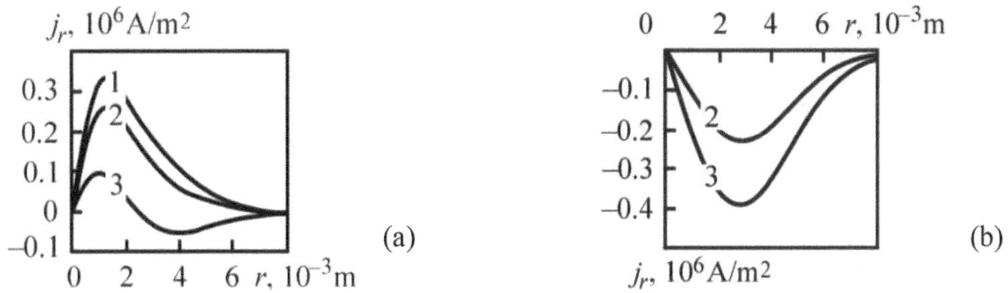

Figure 7. Radial component of current density in the near-anode zone of the arc column: a – calculation with $\varphi_{pa}(r)$ dependence; b – calculation at φ_{pa}= 2.62 V – 1 – $z = L$; 2 – L – 0.25 mm; 3 – L – 0.5 mm.

6 CONCLUSIONS

Verification of the described unified model in the case of argon arc with water-cooled anode leads to the conclusion of its applicability for numerical study of the arc plasma characteristics and characteristics of its impact on the anode surface. Important advantage of this model compared to the known ones is taking into account the non-uniformity of the anode voltage fall distribution in the region of anode binding of the arc, which allows a more adequate description of the processes of energy and charge transfer between the arc plasma and the anode surface.

ACKNOWLEDGMENTS

Supported by GDAS'Project of Science and Technology Development (2020GDASYL-20200301001); The National Key Research and Development Program of China (grant number 2020YFE0205300); Science and Technology Planning Project of Guangdong Province (2022B1212050001); Scientific Research Project:(G2022030063L).

REFERENCES

Almeida, N.A., Benilov, M.S. & Naidis, G.V. 2008. Unified modelling of near-cathode plasma layers in high-pressure arc discharges. *Journal of Physics D: Applied Physics*, 41(24): 245201.

Cressault, Y., Murphy, A.B., Teulet, P., Gleizes, A. & Schnick, M. 2013. Thermal plasma properties for Ar-Cu, Ar-Fe and Ar-Al mixtures used in welding plasmas processes: II. Transport coefficients at atmospheric pressure. *Journal of Physics D: Applied Physics*, 46(41): 415207.

Dinulescu, H.A. & Pfender, E. 1980. Analysis of the anode boundary layer of high intensity arcs. *Journal of Applied Physics*, 51(6): 3149–3157.

Essoltani, A., Proulx, P., Boulos, M.I. & Gleizes, A. 1994. Volumetric emission of argon plasmas in the presence of vapors of Fe, Si, and Al. *Plasma Chemistry and Plasma Processing*, 14: 437–450.

Goodarzi, M., Choo, R. & Toguri, J. M. 1997. The effect of the cathode tip angle on the GTAW arc and weld pool: I. Mathematical model of the arc. *Journal of Physics D: Applied Physics*, 30(19): 2744.

Hsu, K.C., Etemadi, K. & Pfender E. 1983. Study of the free-burning high-intensity argon arc. *J. of Appl. Phys.* 54(3): 1293–1301.

Krivtsun, I., Demchenko, V., Krikent, I., Reisgen, U., Mokrov, O. & Zabirov A. 2015. Distributed and integrated characteristics of the near-anode plasma of the welding arc in TIG and Hybrid (TIG + CO_2 Laser) welding. *Mathematical Modelling of Weld Phenomena*, 11: 837–874.

Mougenot, J., Gonzalez, J.J., Freton, P. & Masquère, M. 2013. Plasma–weld pool interaction in tungsten inert-gas configuration. *Journal of Physics D: Applied Physics*, 46(13):135206.

Nestor, O.H. 1962. Heat intensity and current density distributions at the anode of high current, inert gas arcs. *J. of Appl. Phys.* 33(5): 1638–1648.

Sanders, N.A. & Pfender, E. 1984. Measurement of anode falls and anode heat transfer in atmospheric pressure high intensity arcs. *Journal of Applied Physics*, 55(3): 714–722.

Semenov, I.L., Krivtsun, I.V. & Reisgen, U. 2016. Numerical study of the anode boundary layer in atmospheric pressure arc discharges. *Journal of Physics D: Applied Physics*, 49(10): 105204.

Tanaka, M. & Ushio, M. 1999. Observations of the anode boundary layer in free-burning argon arcs. *Journal of Physics D: Applied Physics*, 32(8): 906–912.

Wendelstorf, J., Simon, G., Decker, I. & Wohlfahrt, H. 1997. Investigation of cathode spot behaviour of atmospheric argon arcs by mathematical modeling. *Proc. 12th Int. Conf. Gas Discharges & Their Applications, Greifswald*, 1: 62–65.

Welding and Related Technologies – Krivtsun et al. (Eds)

An analytical mathematical model for melting layer thickness of HDPE pipes at thermal butt-fusion welding

Donghu Zeng, M. Iurzhenko & V. Demchenko
Department of Plastics Welding at E.O. Paton Electric Welding Institute of the NAS of Ukraine, Kyiv, Ukraine

ABSTRACT: The proper melting layer thickness ensures adequate material mixing, interface pressure, defect prevention, and strong mechanical properties. The purpose of this study is to develop an analytical mathematical model for predicting melting layer thickness at different heat soaking times using the Neumann moving boundary conditions. To verify the reliability of the analytical results, a 2D numerical model in width was also built in ANSYS Fluent 2020R1. It's found that the developed analytical mathematical model provides a reliable approach for predicting the melting layer thickness in HDPE pipes during the thermal butt-fusion welding. The basic alignment of the predictions with those obtained from the numerical model, with an error within 18%, validating its effectiveness.

INTRODUCTION

High-density polyethylene (HDPE) pipes have made significant contributions to our daily lives (James B. Goddard & LLC, Powell, OH, 2011). Interest in innovative studies on HDPE pipes continues to grow, particularly concerning manufacturing processes, structure-property relationships, and novel applications (Niou et al., 2018; Yu et al., 2017). HDPE pipes are extensively utilized for water transportation in municipal engineering and are increasingly being adopted in nuclear power plants (Bai et al., 2024; Cai et al., 2018; Caleb Frederick, et al., 2009). Most researchers always put their eyes on interests of studies on testing techniques, toughness evaluation, welding methods, aging phenomena, machining operations, safety appraisal, and numerical analysis (Akkurt, A., 2014; Belhadi et al., 2016; Guo et al., 2017; Hamlaoui et al., 2017).

The purpose of this study is to develop an analytical mathematical model for predicting the melting layer thickness of HDPE pipes based on heat soaking time using two governing equations of heat transfer in solid- and melting-phase and the interface energy at solid-melting coexistence zone with help of the Neumann moving boundary conditions, while a numerical analysis is also conducted in ANSYS Fluent 2020R1 for validating reliability of results from the analytical mathematical model.

2 MATERIALS AND METHODS

2.1 *Material properties of HDPE pipes*

The material properties of HDPE pipes are non-linear and temperature dependent, where thermal conductivity, isobaric heat capacity, and enthalpy. The initial temperature of HDPE pipes (T_0) was set to be 28°C, while the temperature of heated plate (T_g) was set to be at 240°C. The material properties of HDPE pipes came from the existing literature (Cai et al., 2018).

2.2 *Analytical mathematical model for solid-melting phase*

The heat soaking stage is one of the most critical stages in thermal butt-fusion welding, as it ensures that the entire welding interface reaches a uniform temperature, the high quality of

DOI: 10.1201/9781003518518-23

welded joints in HDPE pipes can be determined by the reasonable heating time and heating temperature. There are the solid phase and the melting phase occurring in HDPE pipes during thermal butt-fusion welding. Additionally, the interface boundary of the solid-melting coexistence zone continuously move as the heat soaking time increases. To predict the melting layer thickness of HDPE pipes at different heat soaking time, this study developed appropriate analytical mathematical model employing the Neumann Solutions. In this study, the following assumptions should be taken for simplifying the complexity of the analytical mathematical model: (a) neglecting the effects of convective heat transfer, (b) the density at the solid-melting phase interface is equal to average density of solid and melting, (c) initial temperature of the HDPE pipes, T_0.

Two governing equations for solid- and melting-phase heat transfer are given by eq. (1)

$$\begin{cases} \dfrac{\partial^2 T_l}{\partial x_l^2} = \dfrac{1}{\alpha_l}\dfrac{\partial T_l}{\partial t}\ for\ meltingphase, x = 0 \Rightarrow T_l{=}T_g; t = 0 \Rightarrow T_l{=}T_0 \\ \dfrac{\partial^2 T_s}{\partial x_s^2} = \dfrac{1}{\alpha_s}\dfrac{\partial T_s}{\partial t}\ for\ solidphase, x \to \infty \Rightarrow T_s{=}T_0; t = 0 \Rightarrow T_s{=}T_0 \end{cases} \tag{1}$$

where suffix s and l donate the solid phase and melting phase, α stands for thermal diffusivity.

At solid-melting coexistence zone, the interface energy can be described by following equation

$$k_l \frac{\partial T_l(x_l, t)}{\partial x} - k_s \frac{\partial T_s(x_s, t)}{\partial x} = \frac{\rho_l + \rho_s}{2} L \frac{dx(t)}{dt} \tag{2}$$

The continuity of temperature at solid-melting coexistence zone should be considered, the mathematical description is given

$$T_l(x_l, t) = T_s(x_s, t) = T_m \tag{3}$$

To obtain an analytical mathematical model for the location of the solid-melting coexistence zone, an error function is used in this study

$$\begin{aligned} T_l(x_l, t) &= T_g + \frac{T_m - T_g}{erf(\gamma)} erf\left(\frac{x_l}{\sqrt{4\alpha_l t}}\right) \\ T_s(x_s, t) &= T_0 + \frac{T_m - T_0}{erfc(\gamma\sqrt{\alpha_l/\alpha_s})} erfc\left(\frac{x_s}{\sqrt{4\alpha_s t}}\right) \end{aligned} \tag{4}$$

where γ is given by

$$\frac{e^{-\gamma^2}}{erf(\gamma)} - \sqrt{\frac{\alpha_l}{\alpha_s}} \frac{k_s}{k_l} \frac{T_m - T_0}{T_m - T_g} \frac{e^{-\gamma^2(\alpha_l/\alpha_s)}}{erfc(\gamma\sqrt{\alpha_l/\alpha_s})} = \frac{\gamma L \sqrt{\pi}}{c_l(T_m - T_g)} \tag{5}$$

γ can be solved employing MATLAB, 0.4501.

Thus, the relationship of melting layer thickness of HDPE pipes and heat soaking time can be derived

$$x_l = 2.9032 \times 10^{-4}\sqrt{t} \tag{6}$$

Theoretically, under a heating temperature of 240°C, the melting layer thickness of HDPE pipes at any time can be calculated using Equation (6).

2.3 *Numerical model of HDPE pipes*

To simulate the temperature distribution and mass fraction of HDPE pipes during thermal butt-fusion welding, this study established two-dimensional and transient computational fluid dynamics (CFD) models with dimensions of 50 mm in length and 11.8 mm in width using ANSYS Fluent 2020 R1, and the element size of mesh is 0.2mm with 15060 nodes and 29500 elements.

For the numerical model scheme based on the Pressure-Based method, the SIMPLE algorithm was used as the solution method, and the QUICK scheme was chosen for numerical discretization. The convergence criteria were 10^{-3} for continuity and 10^{-6} for energy. The Laminar model for viscosity was selected to examine the fluid behaviour because the Reynolds number was less than 2200, and the Solidification and Melting model was enabled. The Under-Relaxation-Factor was set to its default value. In this study, the heat soaking time was 180 seconds, the Number of Time Steps was set to be 360, the Time Step Size was set to be 0.5 seconds, and the Max Iterations was set to be 30. The most important aspect of this study is that the heating temperature is prescribed at red edge of 2D numerical model, which is 240°C.

3 RESULTS AND DISCUSSIONS

3.1 *Temperature distribution analysis*

For investigating the effects of heat soaking time on HDPE pipes, in this study, the temperature distribution of two-dimensional HDPE pipes under heating temperature of 240°C was analysed using ANSYS Fluent 2020 R1. It was found that the temperature distribution exhibits a water ripple pattern from the upper right to the lower left as shown in Figure 1(a). This temperature distribution pattern is caused by different convective heat transfer coefficient between the outer and inner surfaces resulting in different heat transfer capabilities. Temperature contours become denser in regions with large temperature fluctuations and sparser in regions with small temperature changes. Heat conduction in HDPE pipes occurs from the heated end to the non-heated end. It is crucial for technical engineers to understand how temperature is distributed on HDPE pipes at different heating temperatures and heat soaking times, as this knowledge can help determine the optimal welding parameters for achieving the best welded joints.

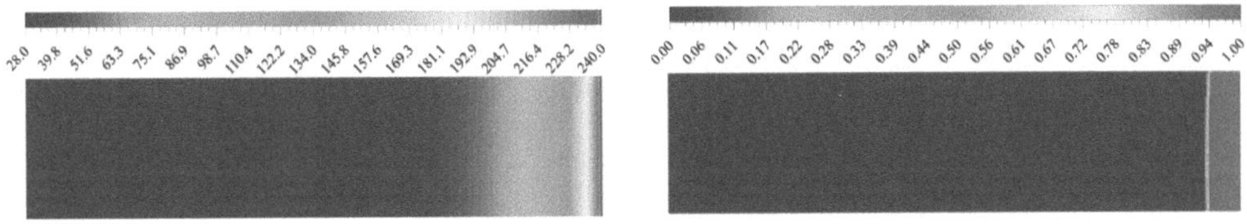

Figure 1. Numerical results at t = 180s, a. temperature distribution (left) and b. mass fraction (right).

3.2 *Melting layer thickness analysis*

This study conducted an analytical mathematical and numerical analysis of melting layer thickness of HDPE pipes at various heat soaking times when the end was heated by the heated plate with 240°C. The mass fraction of HDPE pipes during soaking is shown in Figure 1(b). The blue zone represents the non-fusion zone, while the red zone represents the fusion zone. It was found that the solid-melting coupling zone (indicated by the colour gradient from red to blue) follows an "S" pattern. The mass fractions at the interface among the air phase, melting phase, and solid phase on the outer and inner surfaces differ slightly. The convective heat transfer coefficient at the outer surface is higher than at the inner surface, which is why the melting layer on the outer surface is thinner than on the inner surface. The thickness of the solid-melting coupling zone is approximately 0.3 mm. The melting layer thickness of HDPE pipes increases with heat soaking time, as shown by the red curve in Figure 2.

By substituting heat soaking times into Equation (6), the melting layer thickness of HDPE pipes can be determined through the analytical mathematical model. As observed in Figure 2, the analytical and numerical results are roughly the same. However, the numerical results are consistently smaller than the analytical results due to the neglect of convective heat transfer effects between the surface of the HDPE pipes and the environment but accounting for these effects in the numerical model. Thus, the difference between the numerical and analytical results is reasonable. By comparing the melting layer thickness from the analytical mathematical model to the numerical model, it was found that the error between them is within 18%.

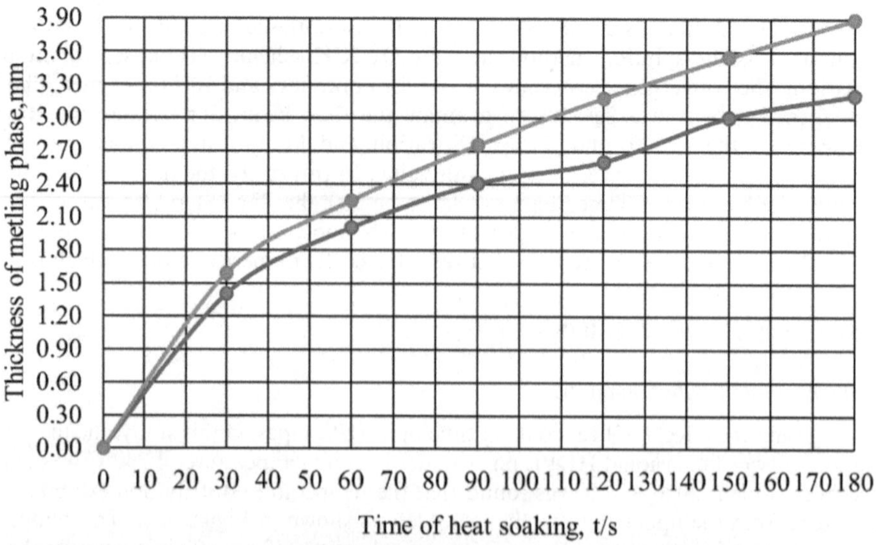

Blue curve and red curve denote analytical mathematical model and numerical model, respectively.
Figure 2. Melting layer thickness of HDPE pipes.

4 CONCLUSIONS

This study developed an analytical model to predict the melting layer thickness of HDPE pipes as the heat soaking time changes during thermal butt-fusion welding.

1. A numerical analysis using ANSYS Fluent 2020R1 examined the temperature distribution and mass fraction of HDPE pipes under the same material properties and welding conditions.
2. An analytical model, based on Neumann's Solution and governing equations for the solid and melting phases, predicted the melting layer thickness and the relationship between the interface boundary location and heat soaking time.
3. The analytical and numerical models produced similar results, with numerical results being slightly smaller and an error margin of 18%.

REFERENCES

Shi, J., et al. 2009. A Model for Predicting Temperature of Electrofusion Joints for Polyethylene Pipes. *Journal of Pressure Vessel Technology* 131(6): 061403 (8 pages).

Akkurt, A. 2014. An analysis of electro-melting and hot element welding method' safety to join PE natural gas pipes. *International Journal of Mechanical & Mechatronics Engineering* 3(2): 493–504.

Bai, C., et al. 2024. Investigation of Creep Behavior of HDPE Pipe Butt Fusion Welded Joints Using a Stepped Isostress Method. *Polymers* 16: 1803. https://doi.org/10.3390/polym16131803

Belhadi, S., et al. 2016. Machining Optimization of HDPE Pipe Using the Taguchi Method and Grey Relational Analysis. *International Polymer Processing* 31, 491–502.

Cai, Z., et al. 2018. Investigation on the hot melting temperature field simulation of HDPE water supply pipeline in gymnasium pool. *Results in Physics* 9: 1050–1056.

James B. Goddard & LLC, Powell, OH. 2011. A Brief History of the Development and Growth of the Corrugated Polyethylene Pipe Industry in North America.

Guo, S., et al. 2017. Safety assessment of high-density polyethylene pipe with thermal damages. *Plastics, Rubber and Composites* 46(4): 173–183.

Hamlaoui, N., et al. 2017. Machining of tough polyethylene pipe material: surface roughness and cutting temperature optimization. *Int J Adv Manuf Technol* 92: 2231–2245.

Caleb Frederick, et al. 2009. High-Density Polyethylene Piping Butt-Fusion Joint Examination Using Ultrasonic Phased Array. *Journal of Pressure Vessel Technology* 132(5): 051501 (8 pages).

Niou, S., et al. 2018. A method for mechanical property assessment across butt fusion welded polyethylene pipes. *Int J Adv Manuf Technol* 97: 543–561. https://doi.org/10.1007/s00170-018-1908-y

Yu, K., et al. 2017. A review of the design and analysis of reinforced thermoplastic pipes for offshore applications. *Journal of Reinforced Plastics and Composites* 36: 1514–1530.

Welding and Related Technologies – Krivtsun et al. (Eds)

Calculation prediction of welding distortions of large-sized cylindrical structures

O.V. Makhnenko, O.S. Milenin, O.A. Velikoivanenko, G.P. Rozynka & O.M. Savytska
E.O. Paton Electric Welding Institute NAS of Ukraine, Ukraine

ABSTRACT: A mathematical models and calculation algorithms have been developed based on the approximate inherent strain method within the theory of elasticity, which makes it possible to quickly predict the overall distortions of a large-sized cylindrical structure with a number of welded joints. The results of mathematical modeling of distortions of the a large-sized cylindrical vessel (diameter 3200 mm, length 12 m, wall thickness 10 mm) made of aluminum alloy when using the Friction Stir welding process and traditional Arc welding are compared.

1 INTRODUCTION

Large-sized cylindrical structures are typical for manufacturing of pressure vessels, propellant tanks and other industrial equipment. Welding technology is widely used to produce assembly joints of large cylindrical structures, etc (Poklyatskyi et al. 2023), (Vasanthakumar & Sekar 2020). The choice of welding process and welding conditions can significantly affect the accuracy, strength and service life of such welded structures. Mathematical modeling can be effectively used in the technological preparation of manufacturing large cylindrical vessels for in-process prediction of stress-strain state. The calculation prediction of the overall distortions and residual stresses of large structures with a number of welded joints is a particularly difficult problem (Makhnenko & Muzghichenko 2007). The solution of such problem requires large computing resources and a long time for computation, which complicates the use of calculation methods in practice. Thus, the problem of developing effective calculation methods for in-process prediction of stress-strain state during assembly welding of large cylindrical structures is quite relevant.

2 PREDICTION OF DISTORTIONS OF THE CYLINDRICAL SHELL BY INHERENT STRAIN METHOD

To solve the problem of predicting distortions in large cylindrical vessels with a number of welds (Figure 1) a procedure of modelling welding stresses and strains based on the combined application of the thermoplasticity analysis (Makhnenko 1976) and approximate inherent strain method was used (Makhnenko 2010). By means of the thermoplasticity analysis temporary and residual stresses and strains for typical welded joints were modelled. It is known that residual welding stresses and plastic strains are formed in a local area in the welded joint zone. Therefore, the distribution of residual stresses and inherent strain function can be obtained on simplified models of weld joints, which requires much smaller computer resources and time for calculation.

Regarding large cylindrical vessels with a number of longitudinal and circumferential butt joints, it is advisable to consider two types of simplified models (Makhnenko & Tsaryk 2024). To determine the local residual plastic strains (inherent strains) in the zones of longitudinal joints, the 3D model of a butt joint of plane elements of a limited size can be used. In the zone of circumferential joints a simplified 2D model of a butt circumferential joint of a real size can

DOI: 10.1201/9781003518518-24

be used, but with the assumption of a fast-moving welding heating source and "plain strain" hypothesis. Then, by means of a 3D model of a cylindrical vessel of a real size using the approximate inherent strain method within elastic formulation of the problem, the overall distortions of large welded structure due to all welds are modelled.

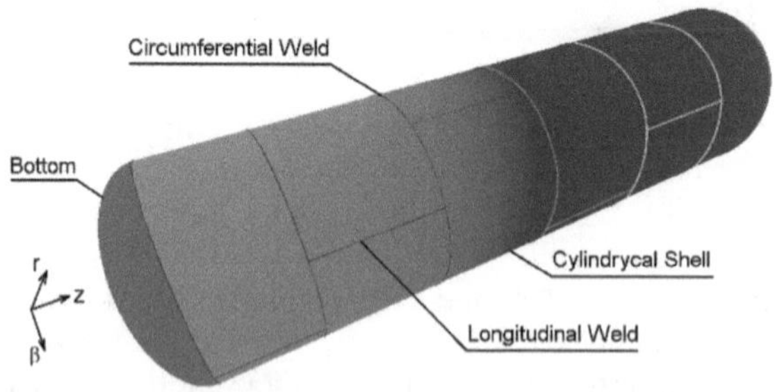

Figure 1. Cylindrical vessel with a number of longitudinal and circumferential welded joints.

The inherent strain function can be presented as a tensor of plastic strains:

$$\Pi_{ij} = \begin{Vmatrix} \varepsilon^p_{zz} & \varepsilon^p_{z\beta} & \varepsilon^p_{zr} \\ \varepsilon^p_{\beta z} & \varepsilon^p_{\beta\beta} & \varepsilon^p_{\beta r} \\ \varepsilon^p_{rz} & \varepsilon^p_{r\beta} & \varepsilon^p_{rr} \end{Vmatrix} = \begin{Vmatrix} \varepsilon^p_{zz} & 0 & 0 \\ 0 & \varepsilon^p_{\beta\beta} & 0 \\ 0 & 0 & -(\varepsilon^p_{zz} + \varepsilon^p_{\beta\beta}) \end{Vmatrix} = \begin{Vmatrix} \Pi_{zz} & 0 & 0 \\ 0 & \Pi_{\beta\beta} & 0 \\ 0 & 0 & -(\Pi_{zz} + \Pi_{\beta\beta}) \end{Vmatrix} \quad (1)$$

where $\varepsilon^p_{r\beta} = \varepsilon^p_{\beta r} = \varepsilon^p_{rz} = \varepsilon^p_{zr} = \varepsilon^p_{z\beta} = \varepsilon^p_{\beta z} = 0$, since the axes Z and β are the main for the longitudinal and transverse directions of welded joints; the condition of volume preserving $\varepsilon^p_{rr} = -(\varepsilon^p_{zz} + \varepsilon^p_{\beta\beta})$. Assuming that an inherent strain function along the welded joint is constant, then $\Pi_{zz} = \varepsilon^p_{zz}(\beta, z)$, $\Pi_{\beta\beta} = \varepsilon^p_{\beta\beta}(\beta, z)$. The tensor of full strains ε_{ij} in the welded shell will be determined by the sum of the tensor of elastic strains ε^e_{ij} and the tensor of inherent strain function Π_{ij}:

$$\varepsilon_{ij} = \varepsilon^e_{ij} + \Pi_{ij}, (i, j = r, \beta, z) \quad (2)$$

The boundary problem for determining welding distortions of the cylindrical shell by inherent strain method can be resolved by analytical way (Makhnenko & Tsaryk 2024) or by using FEM analysis.

3 EXAMPLE OF CALCULATION

The developed mathematical model and calculation algorithm based on the inherent strain method can be effectively used in the technological preparation of the production of large-sized cylindrical structures for operational prediction of the stress-strain state caused by assembly welding. As an example, the assembly welding of a large-sized cylindrical vessel (diameter 3200 mm, total length 12 m, wall thickness 10 mm) made of aluminum alloy is considered. The results of mathematical modeling of welding distortions of the considered structure when using the Friction Stir Welding process (FSW) and traditional Arc Welding are compared.

To determine the inherent strain function, the methods of mathematical modelling of temperature distributions and thermoplasticity analysis of a butt joint of plates at welding heating were used. The results of residual distributions of longitudinal and transverse plastic strains at Arc Welding and FSW were obtained (Figure 2). It should be seen that calculated transverse plastic strains in terms of absolute value are approximately 7-8 times higher than longitudinal, but the zone of longitudinal plastic strain formation is 1.5–2 times wider. It should be also noted that calculated transverse and longitudinal plastic strains for FSW are rather lower then for Arc Welding: near 2 times lower in absolute value and zone of their formation is narrower.

The obtained distributions of residual longitudinal and transverse plastic strains were used in the prediction by the inherent strain method of the overall distortions of the large-sized cylindrical vessel (Figure 1), which contains six longitudinal and five circumferential welded joints (location coordinates of the circumferential welded joints z=1.5, 3, 4.5, 6, 7.5, 9 m, longitudinal welded butt joints at βR=0, ±755 mm). The displacements of the surface points of the cylindrical shell was defined by the finite element method on a full-size model for each welded joint separately and then overall displacements were added by the principle of superposition.

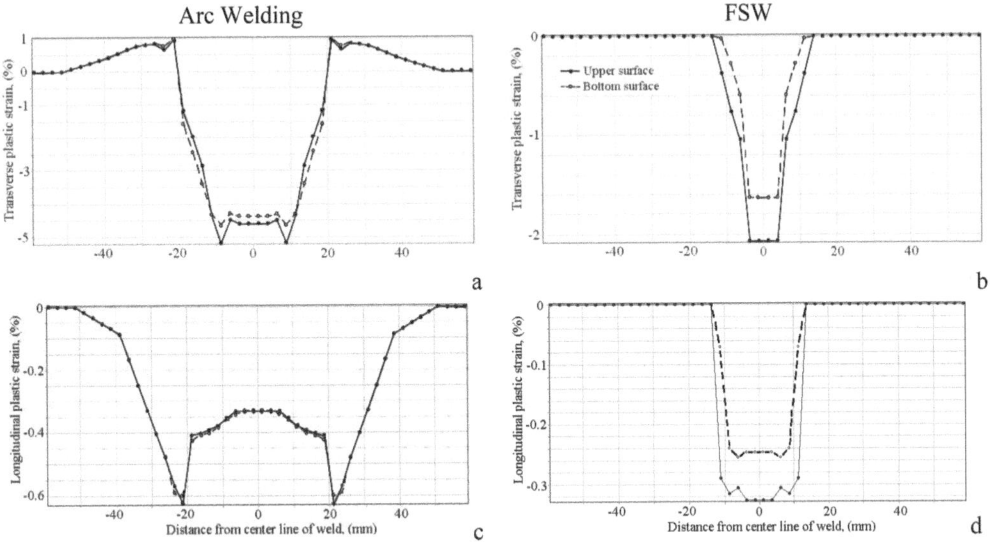

Figure 2. Results of calculating the distribution of plastic strains (inherent strain function) for butt welded joints in the 3D model of plates (300×300 mm, δ=10 mm) for Arc Welding (a, c) and FSW (b, d).

The results of prediction the residual distortions (radial and axial displacements) at final stage of assembly welding of a large-sized cylindrical vessel are presented on Figure 3 and Figure 4. Distortions have a complex nature of distribution, for Arc Welding radial deflections up to $U_r \approx$ -6.5 mm in the zones of circumferential and longitudinal welded joints are changed by shell bends up to $U_r \approx$ 6.0 mm (Figure 4a) in the zones between welded joints. For FSW the level of residual distortions of the shell is noticeably lower, so the radial deflections do not exceed U_r = -2 mm (Figure 4b). Axial displacements (shrinkage of the shell length) for FSW are 5 times lower $U_z \approx$ -1 mm (Figure 4d) than for Arc Welding $U_z \approx$ -5 mm (Figure 4c). Thus, the use of FSW during assembly welding of a large-sized cylindrical vessels made of aluminum alloy allows to significantly reduce the level of residual distortions.

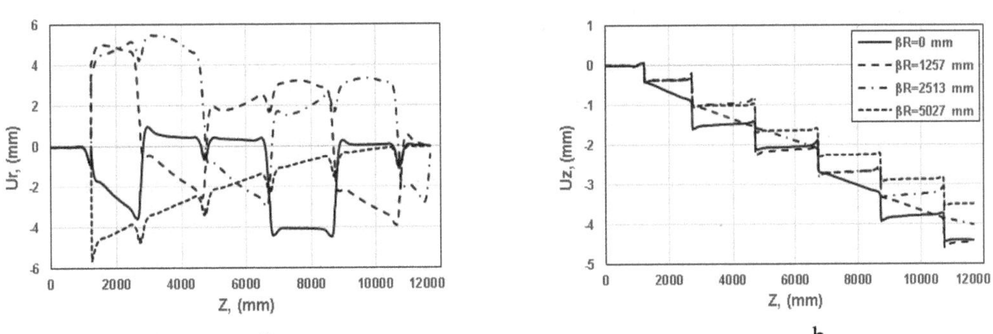

Figure 3. Distribution of radial U_r (a) and axial U_z *(b)* displacements of cylindrical vessel shell after Arc welding of circumferential and longitudinal butt joints.

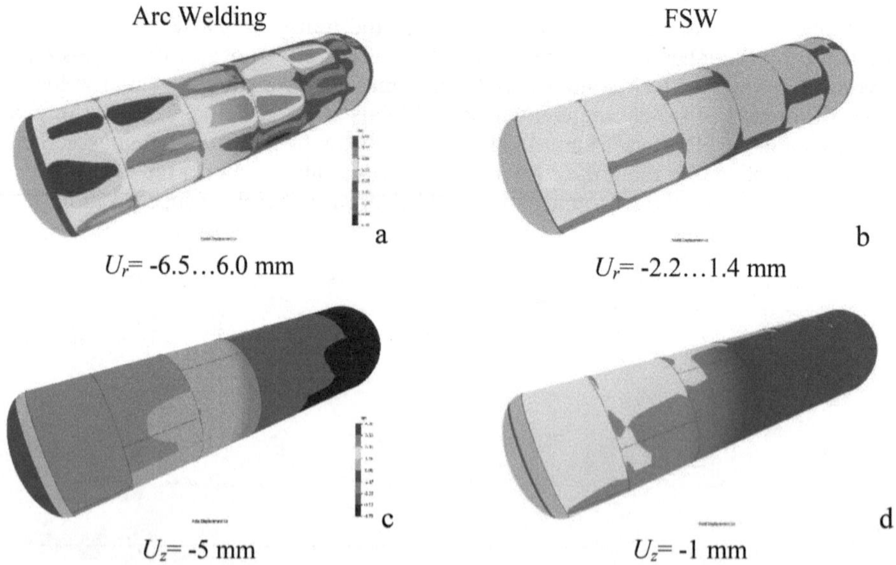

Figure 4. Results of simulation of distortions (radial U_r and axial U_z components) of cylindrical vessel shell (diameter 3200 mm, length 12 m, wall thickness 10 mm) after welding of circumferential and longitudinal butt joints by Arc Welding (a, c) and FSW (b, d).

4 CONCLUSIONS

1. The mathematical models and calculation algorithms for operative numerical determination of welding strains of large-sized cylindrical shells with a number of circumferential and longitudinal welded joints have been developed based on the inherent strain method.
2. The inherent strains of welded joints of cylindrical shell (plastic strains in the axial Π_{zz} and circumferential $\Pi_{\beta\beta}$ directions) can be determined by thermoplasticity analysis on simplified models of longitudinal and circumferential welded joints.
3. The developed specialized software can be effectively used for in-process prediction of a stress-strain state in assembly welding of large cylindrical vessels.

REFERENCES

Makhnenko O.V. & Muzghichenko A.F. 2007. Mathematical modeling of thermal straightening of cylindrical shells and shafts with general distortions of curvature of the longitudinal axis. *Avtomat. Svarka* 9: 23–28

Makhnenko O.V. 2010. Combined use of the method of thermoplasticity and the method of the shrinkage function for the study of the process of thermal straightening of shipbuilding panels. *Journal of Mathematical Sciences* 167(2): 232–241

Makhnenko O.V. & Tsaryk B.R. 2024. Mathematical modeling of distortions at welding of large vessels of aluminum alloy. *The Paton Welding Journal* 08:17–24

Makhnenko V.I. 1976. *Calculation methods for investigation kinetics of welding stresses and deformations.* Kyiv: Naukova dumka

Poklyatskyi A.G., Motrunicc S.I., Fedorchuk V.Ye., Falchenko Iu.V., Sagul M.. 2023. Mechanical properties and structural features of butt joints produced at FSW of aluminium alloys of different alloying systems. *The Paton Welding Journal 4*: 3–10

Vasanthakumar Pandian & Sekar Kannan 2020. *Numerical prediction and experimental investigation of aerospace-grade dissimilar aluminium alloy by friction stir welding.* Journal of Manufacturing Processes 54: 99–108

Welding and Related Technologies – Krivtsun et al. (Eds)

Mathematical modeling of heat and mass transfer processes of electron beam melting of ingots from high-strength titanium alloys

S.V. Akhonin, I.V. Krivtsun, V.O. Berezos, S.V. Rymar & R.S. Hubatiuk
E.O. Paton Institute of Electric Welding of the National Academy of Sciences of Ukraine

ABSTRACT: A mathematical model of temperature fields and laminar hydrodynamic flows during electron-beam melting of high-strength titanium alloy was developed, which made it possible to determine the geometry of the liquidus-solidus interphase (mushy) zone. The zone determines the crystallization front of the metal whose position and geometry significantly affects the quality of the ingot formation. According to the calculated regimes at E.O. Paton Electric Welding Institute, high-quality ingots of TA15 titanium alloy were smelted a 150 mm diameter.

1 INTRODUCTION

Knowledge of the processes of heat and mass transfer in high-temperature melts of titanium alloys during their electron beam melting (EBM) and the formation of ingots is important, which is an urgent task of modern metallurgy, and the process of metal crystallization is one of the key factors in the smelting of high-quality ingots. Uniform distribution of alloying elements in the ingot is ensured in the case of a flatter crystallization front with uniform temperature distribution in the ingot, which is a prerequisite for the homogeneity of the solid metal structure in it.

The studies were carried out for a 150 mm diameter ingot, which is formed in a continuous crystallizer with an ingot extraction speed of 40 kg/h. At the same time, the movement of liquid metal flows has a laminar character. Similar studies for a given diameter of the crystallizer and the speed of drawing the TA15 alloy ingot (Ti-6.5Al-2Zr-1Mo-1V) at EBM are being conducted for the first time.

2 PROBLEM STATEMENT

With the development of computer technologies and the appearance of powerful multiphysics computer programs, the possibility of sufficiently accurate numerical simulation of heat and mass transfer problems in metallurgy has appeared (Haag et al. 2020, Ravindra Pardeshi 2016, Ufodike et al. 2023, Vanja Hatič et al.2020, Xiaowei Xu et al. 2022, Yajun Luo & Zhifeng Zhang 2019), which has been proved by numerous experiments and made the search for metal melting modes cheaper with an increase in its quality in ingots. In the E.O. Paton Electric Welding Institute and IED of the National Academy of Sciences of Ukraine (Akhonin et al 2021, Akhonin et al 2019) mathematical models have been developed that became the basis for solving similar problems. These works considered mathematical modeling of heat and mass transfer processes in ingots with a diameter of 400 and 600 mm for the titanium alloy Ti-6Al-4V when using two electron beams falling on the surface of the molten bath. When solving the task of this work, the mathematical models must undergo modernization taking into account other parameters of melting – a smaller diameter of the crystallizer, a higher speed of pouring metal into it, etc.

DOI: 10.1201/9781003518518-25

3 PURPOSE AND OBJECTIVES OF THE RESEARCH

Mathematical modeling of heat transfer and hydrodynamic processes during melting of a cylindrical ingot with a diameter of 150 mm of titanium alloy TA15 during EBM and determination of the necessary ingot melting modes is the aim of the work. Achieving this goal makes it possible to establish the shape of the metal crystallization front, which affects the quality of ingot formation. In order to achieve this, it is necessary to solve the following tasks: to develop a mathematical model of heat transfer and hydrodynamic processes in the ingot with the laminar nature of the movement of the liquid metal, taking into account its thermophysical characteristics, and to determine the distribution of temperature fields in the ingot and laminar hydrodynamic flows in the molten pool.

4 MATERIALS AND RESEARCH METHODS

The object of research in the work is a mathematical model of laminar processes of heat and mass transfer during EBM of TA15 titanium alloy ingot. The subject of research is the propagation of hydrodynamic flows in the melt and temperature fields in the ingot.

The main research hypothesis is that at a constant rate of pouring molten metal into the crystallizer, the position of the electron beam on the surface of the melt bath and its power are the factors that mainly affect the position of the metal crystallization front and the quality of the metal in the ingot.

When performing the work, assumptions were made:

- the liquid and solid phases of the ingot metal are considered as one region with a liquid phase, and the solid metal corresponds to the volume in which the movement is suppressed by the Darcy braking force (except for the ingot pulling speed) and a fixed calculation grid is used in the method finite elements;
- the interphase boundary of the "liquidus-solidus" alloy L–S is a viscous mushy zone in which the boundary is determined by the liquidus and solidus temperatures, and the metal parameters change according to the proposed law relative to the volume fraction of these phases;
- the volume fraction of the change zone metal versus temperature is a smoothed function $F_{L\text{-}S}$, which smoothly changes from 0 to 1, based on the mathematical error function erf;
- approximate dependences of thermophysical parameters of titanium alloy TA15 with smooth stepwise jumps of values at phase change are used, also with the use of the erf function with artificially expanded (Rubinetti 2018) temperature limits in phase change;
- the distribution of the thermal power of the electron beam on the surface of the bath mirror varies according to the Gaussian law.

The following simplifications are adopted:

- liquid metal is poured inlet the crystallizer continuously at the same speed and a constant rate of ingot extraction is ensured;
- the melt bath mirror is a flat surface; molten metal is taken as an incompressible viscous liquid;
- the heat capacity in the α–β and L–S phase change zones of the alloy takes into account the equivalent heat capacity and the heat capacity from the latent heat of phase change;
- the heat capacity of the latent heat of vaporization of the metal is indirectly taken into account by the specific power of evaporation from the surface.

The mathematical model involves fundamental physical laws of heat transfer, hydrodynamic laws of heat and mass transfer, Navier-Stokes equation. The numerical method of finite elements was used in the calculations. Smelting of ingots took place on the research equipment of the E.O. Paton Electric Welding Institute (Akhonin et al 2022, Akhonin & Pikulin 2019).

5 RESEARCH RESULTS

5.1 *Mathematical model of heat transfer and hydrodynamic processes in a cylindrical ingot with laminar movement of the metal melt*

The main equations used in the model are the following: the heat transfer equation, W/m^3, during laminar motion of liquid metal for a stationary process, taking into account the liquidus-solidus phase change, is written in the form (Akhonin et al. 2019, Heat Transfer Module):

$$\rho C_p \mathbf{u} \cdot \nabla T = \nabla \cdot (\lambda \nabla T); \tag{1}$$

where ρ – density, kg/m^3; C_p – heat capacity, which takes into account the heat of the phase change, J/(kg·K), which can be calculated according to the work formulas (Akhonin et al. 2019); $\mathbf{u}$ – metal velocity vector, m/s; ∇ – nabla operator, m^{-1}; T – temperature, K; λ – thermal conductivity (molecular), W/(m·K). The product $\lambda \nabla T$ is the vector of specific heat flow due to thermal conductivity (heat flow density), W/m^2.

In the mathematical model, heat supply is balanced by heat removal due to radiation and evaporation from the mirror of the metal bath, heat transfer to the crystallizer, radiation from the surface of the ingot into the crystallizer below the point of detachment of the ingot from the walls of the crystallizer, radiation outside the crystallizer, thermal conductivity of the ingot at the lower limit of the calculation area, which are taken into account it is possible to use the equation of the boundary conditions written in (Akhonin et al. 2019).

Mathematical modeling of hydrodynamic processes during the continuous supply of incompressible titanium melt inlet the crystallizer was performed using the model of the Navier-Stokes equations for laminar flow. With a stationary process, the model has the form (Akhonin et al. 2021, Versteeg & Malalasekera 2007):

$$\rho(\mathbf{u} \cdot \nabla)\mathbf{u} = -\nabla \cdot p\mathbf{I} + \nabla \cdot \mu[\nabla \cdot \mathbf{u} + (\nabla \cdot \mathbf{u})^T] + \mathbf{F}_d + \rho \mathbf{g}; \tag{2}$$

$$\rho \nabla \cdot \mathbf{u} = 0. \tag{3}$$

Here p – average pressure value, Pa; $\mathbf{I}$ – a unit tensor; $\mathbf{F}_d$ – the Darcy braking force vector, which is introduced artificially to reduce and suppress velocities in the viscous and solid zones, Pa/m; $\mathbf{g}$ – free fall acceleration vector, m/s^2; μ – the coefficient of molecular viscosity of the liquid metal, Pa·s.

In equation (2), in the last term, ρ is a function of temperature T, and the idea of the Boussinesq approximation works here, in which the product $\boldsymbol{\rho}\mathbf{g}$ is applied to the convection process when T changes, since g is a sufficient value for this product to be significant.

The braking force of Darcy, Pa/m, is calculated according to the formula from works (Avnaim et al. 2016, Ben-David et al. 2013). Suppression of motion in solid metal occurs when $F_{L\text{–}S} = 0$ (except for the melting rate of the ingot $\mathbf{u}_{cast}$). In the melt, when $F_{L\text{–}S} = 1$, the Darcy braking force $\mathbf{F}_d = 0$, and does not affect the movement (Avnaim et al. 2016). In the two-phase zone, the flow corresponds to the Darcy force.

For the hydrodynamic problem, it is also possible to use the equations of the boundary conditions from (Akhonin et al. 2019).

When calculating heat transfer and hydrodynamic processes with an existing phase change in metal, the numerical method of skin elements with a fixed calculation sieve is used, in which the liquid and solid phases are considered as one area. The interphase region is taken into account as a viscous mushy zone, the parameters of which change only in volume fractions of the phase according to a certain law.

The liquid phase is present at a temperature above the liquidus temperature T_L, and the solid phase is present at a temperature below the solidus temperature T_S. Between the temperatures T_L and T_S there is a mushy zone.

In the two-phase zone of the liquidus – solidus change, the $F_{L\text{–}S}$ curve has an increasing dependence, and $(1 - F_{L\text{–}S})$ has a decreasing dependence. By analogy with the $F_{L\text{–}S}$ function, the dependence of the $F_{\alpha\text{–}\beta}$ value in the corresponding temperature range of the α–β phase change was obtained.

5.2 *Results of three-dimensional mathematical modeling*

According to the developed mathematical models of heat transfer and laminar hydrodynamic processes in the ingot, as a result of three-dimensional numerical simulation with the use of temperature-dependent thermophysical characteristics of the metal and interphase mushy zones, temperature fields and hydrodynamic flows were calculated during the production of the TA15 titanium alloy ingot by the EBM method.

Mathematical modeling was carried out for laminar hydrodynamic processes and temperature fields in the ingot in a continuous crystallizer with an internal diameter of 150 mm, a height of 220 mm, and an ingot extraction speed of 40 kg/h. The pouring of pre-molten metal from the cold hearth onto the surface of the bath takes place near the wall of the crystallizer in a square spot of 10 x 10 mm. The surface of the liquid metal bath is heated by an electron beam with a power of 21 kW with an efficiency of 75%, which moves in a circle around the periphery of the bath with a Gaussian distribution of the thermal power from the electron beam.

Figure 1, a shows the estimated three-dimensional distribution of the temperature field and the directions of the velocity vectors of metal flows in the crystallizer with the deduction of the ingot pulling speed to the bottom, Figure 1b shows their top view.

Figure 2a shows the directions of the velocity vectors of the liquid metal flows with the deduction of the ingot pulling speed downwards in the axial direction in a section along its vertical axis of symmetry. The size of the vectors is given in a logarithmic scale, proportional to the speed of the melt movement.

Figures 2a illustrates that the flow of liquid metal spreads from the place of its pouring from one side to the walls of the crystallizer, to the opposite side of the wall, forming a small vortex near the latter. The figure shows the isotherms of liquidus T_L (upper curve) and solidus T_S (lower curve) at the boundary of the crystallization front. The position and geometry of this front (Figures 1a, 2a) largely determines the quality of the ingot formation and the concentration of the distribution of alloying elements in its volume. When viewed from above (Figure 1b), a pronounced diametrical distribution of metal flows is observed, which is caused by the action of flows from liquid metal pouring into the crystallizer and thermal power from the electron beam. The results of calculations of melt movement trajectories showed an adequate effect of the Darcy function to suppress the movement of metal in the solid phase.

Figure 2b shows a macroslice of the ingot template, cut in the axial plane, on which the crystallization profiles of the metal are clearly visible when the melt is poured into the crystallizer from the cold hearth. A comparison of the contours of the crystallization fronts of the ingot, one of which is shown on the template in Figure 2b with a light solid line and the calculated contour of the crystallization front, which is marked with a broken line, indicates a good coincidence of the profiles of the fronts and acceptable accuracy of the results of mathematical modeling.

The macrocut in Figures 2b is characterized by the absence of shells and discontinuities, and the macrostructure of the ingot has crystals close to equiaxed ones, which is an indicator of good ingot quality.

An important indicator of the quality of ingots of titanium alloys is the quality of their side surface. In order to reduce corrugations on the side surfaces of ingots during their melting, it is necessary to strive for the liquid metal of the pool to be less in contact with the wall of the crystallizer. This is ensured by the choice of the position of the electron beam on the surface of the pool and its intensity. The same factors affect the geometry of the metal crystallization front. Mathematical modeling made it possible to obtain the necessary melting parameters to ensure these conditions. In the ingot of titanium alloy TA15 with a diameter of 150 mm and a length of 500 mm, obtained by the EBM method in accordance with the calculated modes, the side surface does not contain defects of casting origin, only small corrugations and roughness are observed on it. At the current level of development of the melting of titanium alloys, such ingots are of good quality. In the future, the ingot undergoes mechanical processing and is considered ready for the manufacture of the necessary products from it.

As a result of the numerical simulation, it was established that the power of the electron beam and the type of its flooring on the surface of the pool significantly affect the heat and mass transfer.

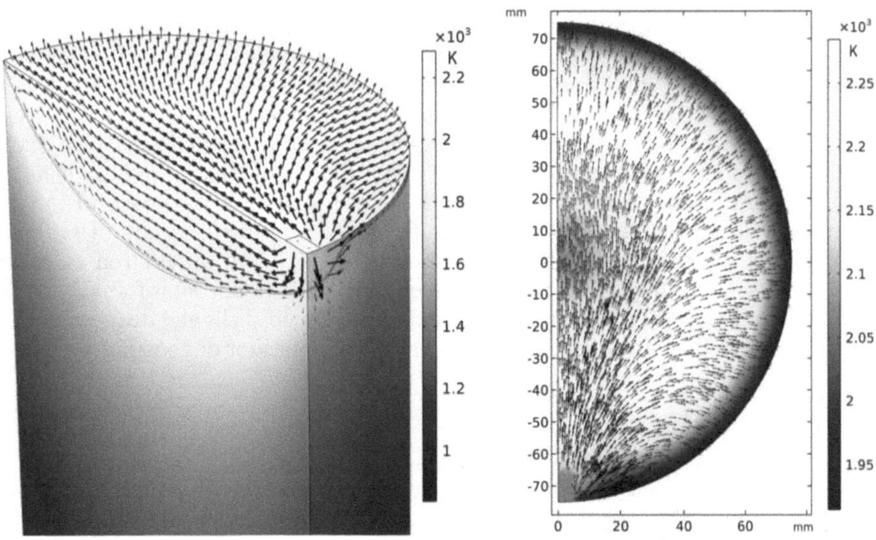

a b

Figure 1. Three-dimensional distribution of the temperature field and metal flows in the crystallizer – a; top view – b.

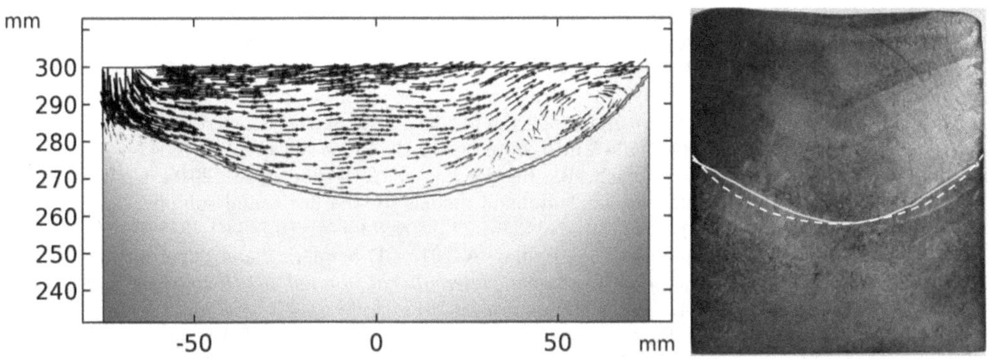

a b

Figure 2. Flows of liquid metal in the bath in a section along the vertical axis of symmetry of the ingot – a; of the macroslices of the ingot template, cut in the axial plane – b, with the calculated crystallization front superimposed on it and the crystallization front of the ingot indicated by a dashed line.

Mathematical modeling made it possible to eliminate from the preparation for conducting the technological process a very expensive method of samples for determining technological regimes and is useful and relevant for foundry engineers.

6 CONCLUSIONS

A three-dimensional mathematical model of thermal processes in the ingot and stationary laminar hydrodynamic processes based on the Navier-Stokes equations in the melt of a cylindrical ingot was developed. The ingot is melted by the electron beam melting method, in which the beam moves in a circle around the periphery of the melt pool with a Gaussian distribution of the thermal power from the beam.

To correctly solve the problem of calculating the laminar processes of heat and mass transfer in the ingot by the finite element numerical method, the dependence of the thermophysical characteristics of the TA15 titanium alloy on temperature, including in the mushy zones, was applied using the mathematical error function erf. With gradual changes in the values of thermophysical characteristics during phase change, this ensured the continuity of these functions.

Based on the results of mathematical modeling, the temperature fields and laminar hydrodynamic flows of molten metal in a TA15 titanium alloy ingot with a diameter of 150 mm were obtained. The geometry of the liquidus-solidus two-phase zone was determined, which determines the crystallization front of the metal, the position and geometry of which significantly affects the quality of the ingot formation. It was found that the distribution of the power of the electron beam on the surface of the pool has a very significant effect on the heat and mass transfer at a constant rate of pouring molten metal into the crystallizer. With an electron beam heating power of 21 kW and a melting speed of 40 kg/h, such a crystallization front is formed that ensures an almost uniform distribution of alloying elements. There are no shells and discontinuities in the ingot, and the crystals in it are close to equiaxed, which is an indicator of good ingot quality.

REFERENCES

Akhonin, S.V., Berezos, V.O., Bondar, O.I., Glukhenkii, O.I., Goryslavets, Yu.M. & Severin, A.Yu. 2021. Mathematical modeling of hydrodynamic and thermal processes at crystallization of titanium ingots produced by EBM. *Electrometallurgy Today (Sovremennaya Elektrometallurgiya)* 1: 27–34. https://doi.org/10.37434/sem2021.01.03

Akhonin, S.V., Gorislavets, Yu.M., Glukhenkiy, A.I., Berezos, V.A., Bondar, A.I. & Pikulin, A.N. 2019. Modeling hydrodynamic and thermal processes in the mould in cold-hearth electron beam melting. *Electrometallurgy Today (Sovremennaya Elektrometallurgiya)* 4: 9–17. https://doi.org/10.15407/sem2019.04.02

Akhonin, S., Pikulin, O., Berezos, V., Severyn, A., Erokhin, O. & Kryzhanovskyi, V. 2022. Determining the structure and properties of heat-resistant titanium alloys VT3-1 and VT9 obtained by electron-beam melting. *Eastern-European Journal of Enterprise Technologies* 5 (12 (119)): 6–12. https://doi.org/10.15587/1729-4061.2022.265014

Akhonin, S.V. & Pikulin, O.M. 2019. Investigation of Effect of Electron Beam Surface Treatment of Titanium Alloy Ingots on Structure and Properties of Melted Metal. *IOP Conference Series: Materials Science and Engineering* 582 (1): 012047. https://doi.org/10.1088/1757-899x/582/1/012047

Avnaim, M.H., Levy, A., Mikhailovich, B., Ben-David, O. & Azulayet, A. 2016. Comparison of three-dimensional multidomain and single-domain models for the horizontal solidification problem. *Journal of Heat Transfer* (138) 11: 112301-1–112301-11. https://doi.org/10.1115/1.4033700

Ben-David, O., Levy, A., Mikhailovich, B. & Azulay, A. 2013. 3D Numerical and Experimental Study of Gallium Melting in a Rectangular Container. *International Journal of Heat Mass Transfer* 67: 260–271. http://dx.doi.org/10.1016/j.ijheatmasstransfer.2013.07.058

Haag, J., Martens, J., Dussoubs, B., Jardy, A. & Bellot, J.-P. 2020. Analysis of the Thermal Transfers in a VASM Crucible: Electron Beam Melting Experiment and Numerical Simulation. *Metals* 10: 1152. https://doi.org/10.3390/met10091152

Heat Transfer Module User's Guide. Version: Comsol 5.4. Available at http://www.comsol.com

Ravindra Pardeshi. 2016. Computational model for multi alloy casting of aluminum rolling ingots. *Journal of Manufacturing Processes* 21: 23–29. http://dx.doi.org/10.1016/j.jmapro.2015.10.006

Rubinetti, D., Weiss, D.A., Chaudhuri, A. & Kraniotis, D. (2018). Modeling Approach to Facilitate Thermal Energy Management in Buildings with Phase Change Materials. *Conference: Comsol Conference 2018: Lausanne, Switzerland*, 7. https://www.comsol.com/paper/modeling-approach-to-facilitate-thermal-energy-management-with-phase-change-mate-63481

Ufodike, C.O., Nzebuka, G.C. & Egole, C.P. 2023. Prediction of Limiting Casting Speed in a Horizontal Direct-Chill Casting through Numerical Modeling and Simulation. *Metals* 13: 1071. https://doi.org/10.3390/met13061071

Vanja Hatič, Boštjan Mavrič & Božidar Šarler. 2020. Simulation of macrosegregation in direct-chill casting – A model based on meshless diffuse approximate method. *Engineering Analysis with Boundary Elements* 113: 191–203. https://doi.org/10.1016/j.enganabound.2019.12.006

Versteeg, H.K. & Malalasekera, W. 2007. An Introduction to Computational Fluid Dynamics. The Finite Volume Method. *Harlow: Pearson Prentice Hall*: 503.

Xiaowei Xu, Jun Li, Zihui Dong, Long Zeng, Zhenming Xu & Jianguo Li, 2022. Multiphase modelling of the continuous metallurgical purification process for impurity removing of recycled aluminum. *Journal of Materials Research and Technology* 18: 830–840. https://doi.org/10.1016/j.jmrt.2022.03.003

Yajun Luo & Zhifeng Zhang. 2019. Numerical modeling of annular electromagnetic stirring with intercooling in direct chill casting of 7005 aluminum alloy billet. *Progress in Natural Science: Materials International*. 29: 81–87. https://doi.org/10.1016/j.pnsc.2019.01.007

Welding and Related Technologies – Krivtsun et al. (Eds)

Finite element modelling of the influence of the filler wire speed on the weld formation during the narrow gap welding of titanium alloys

R.V. Selin, V. Bilous & S.V. Akhonin
E.O. Paton Electric Welding Institute of the NAS of Ukraine, Kyiv, Ukraine

Y. Zhang, B. Qin & H. Wang
China-Ukraine Institute of Welding, Guangdong Academy of Sciences, Guangdong Provincial Key Laboratory of Advanced Welding Technology, Guangzhou, China

ABSTRACT: In narrow gap TIG welding, the speed at which filler wire is fed into the weld pool is a critical parameter. It can significantly influence the weld quality, including the melting of the side walls of the joint, which can lead to a decrease heat input and reduced penetration into the side walls hence increasing the chances of such defects as lack of fusion, undercuts, or cold laps. In this work the finite element model of the narrow gap TIG welding was created. Based in that model, the influence of the filler wire speed on the heat distribution, side wall fusion and weld formation are shown.

1 INTRODUCTION

Narrow gap welding of titanium alloys is a specialized welding technique used to join thick sections of titanium efficiently and with minimal distortion. It involves a narrow weld groove, offers significant advantages in terms of reducing weld volume and minimizing distortion, making it particularly suitable for thick sections of titanium alloys (Näsström et al. 2020).

The Paton Electric Welding Institute has been at the forefront of innovative welding research, including the development of narrow gap welding techniques for titanium alloys under the influence of controlled magnetic fields. This combination of welding and magnetic field control is aimed at addressing several challenges associated with welding titanium, particularly in terms of improving weld quality, stability, and overall mechanical performance (Akhonin et al. 2023).

In narrow gap welding of titanium alloys, the introduction of a controlled magnetic field offers enhanced control over the welding process, particularly in terms of arc stability and weld pool behavior (B Paton et al. 2012). The filler wire speed, which dictates the rate at which filler material is introduced into the weld pool, interacts directly with the arc characteristics and the heat distribution in the narrow gap. Therefore, understanding how this parameter is influenced by a controlled magnetic field is crucial for optimizing welding outcomes (Xu et al. 2015). Mathematical modeling of narrow gap welding of titanium alloys, particularly with varying filler wire speeds, provides a robust framework for understanding the complex interactions that occur during the welding process (Fang et al. 2021).

2 METHODOLOGY AND RESEARCH METHODS

For experiment, 200x75x30mm plates with 10mm gap width of titanium alloy Ti-6Al-4V (Pederson 2002) were used. Welding wire 2V was used as filler wire (Akhonin et al. 2023). Geometry of the sample with mesh size is shown in Figure 1. Welding simulation has been investigated on a few different welding modes, shown in Table 1. For all 3 modes, welding speed, arc voltage, welding current remains the same and equals Vw = 8m/h, U = 12V, Iw = 480A.

The model for studying the influence of filler wire speed on the weld formation was developed based on thermal conductivity and boundary conditions for heat transfer to the

DOI: 10.1201/9781003518518-26

environment, along with magnetic field effects to simulate the impact of the magnetic field on arc stability, heat transfer, and weld pool dynamics:

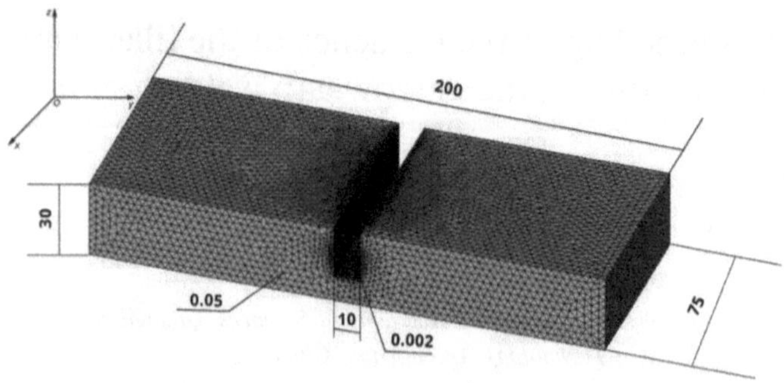

Figure 1. Finite element model of narrow gap TIG welding.

Table 1. Welding modes for magnetically controlled narrow gap TIG welding.

Welding mode	Magnetic induction, mTl	Amplitude, mm	Filler wire speed, m/h
1	-	-	65
2	10	3	65
3	10	5	80

$$\rho c_p \frac{\partial T}{\partial t} = \nabla \cdot (k\nabla T) \tag{1}$$

where T – temperature, K, ρ - is the density of the titanium alloy, kg/m^3, cp - is the specific heat capacity of Ti-6Al-4V alloy, which takes into account the heat of the phase change, J/(kg·K) k - is the thermal conductivity, W/(m*K).

The following boundary conditions describe the heat exchange of the product with the environment (Akhonin et al. 2021):

$$-\lambda \frac{\partial T}{\partial t} = q_{conv} + q_{rad} + q_h \tag{2}$$

Boundary conditions consist from heat loss through convection at the boundaries of the weld pool and workpiece, which is modeled using Newton's law of cooling:

$$q_{conv} = h_{conv}(T - T_\infty), \tag{3}$$

Where h_{conv} is the convective heat transfer coefficient, W/m^{2}K, T is the surface temperature of the workpiece, K, $T\infty$ is the ambient temperature, K.

And Radiative Heat Loss, that is significant at higher temperatures and is modeled using Stefan-Boltzmann's law:

$$q_{rad} = \sigma\left(T^4 - T_\infty^4\right) \tag{4}$$

Where ϵ is the emissivity of titanium, σ is the Stefan-Boltzmann constant, W/m^{2}K^4, T is the surface temperature of the workpiece, K.

Magnetic arc oscillation introduces an oscillatory component to the arc's movement, driven by a magnetic field. In this case, the magnetic field is applied at a frequency of 10 GHz, which causes the arc to oscillate. The force exerted on the weld pool by the magnetic field can be modeled using the Lorentz force. In this case, the magnetic arc oscillation at 10 GHz can be described as an additional sinusoidal movement of the arc. The oscillation of the arc impacts the heat distribution:

Therefore, heat flow distribution in the gap of the model equals:

$$q_h = q_m \exp\left\{-\frac{(y - Asin(\omega t))^2 + (x - Vt)^2}{R^2}\right\} \tag{5}$$

Where q_m - the largest heat flow in the center of the heating spot, J/sec; V – welding speed m/s; R is the radius of the heating spot, m, x, y - coordinates on the surface, A is the amplitude of the oscillation (see Table 1), ω=2π×10 rad/s (oscillation frequency of 10 GHz).

The initial temperature of the workpiece is assumed to be uniform

$$T_{(x,y,z,0)} = T_0 \tag{6}$$

Where T_0 is the ambient temperature (e.g., 298 K).

The speed at which the filler wire is fed into the weld pool affects the overall heat balance because the filler material absorbs heat as it melts. The net heat input is reduced by the energy required to melt the filler wire. The heat balance equation with filler wire speed is:

$$Q_{net} = Q_{arc} - m_{wire} c_p (T_{melt} - T_0) \tag{7}$$

Where m_{wire} - is the mass flow rate of the filler wire, determined by the filler wire speed, kg/s, c_p - is the specific heat capacity of the filler wire 2V, which takes into account the heat of the phase change, J/(kg·K); T_{melt} - is the melting temperature of the titanium alloy, K, T_0 is the ambient temperature, K.

Higher filler wire speeds increase the mass of material being introduced, which requires more energy for melting, reducing the available heat for the weld pool.

3 RESULTS AND DISCUSSION

Finite element modelling allowed to obtain 3 experimental narrow gap welded joints, based on modes from Table 1.

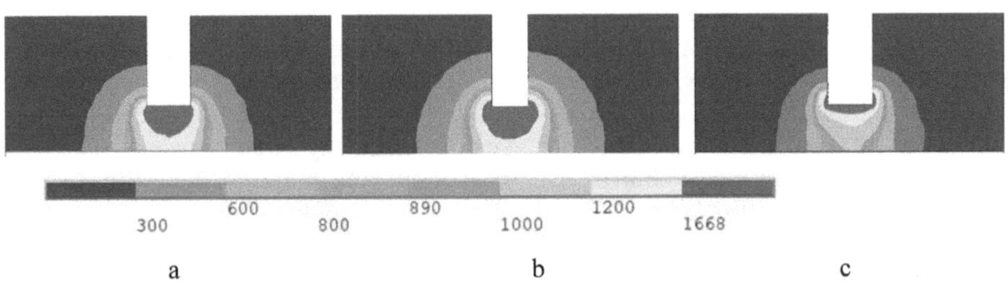

Figure 2. Welds geometry obtained by narrow gap TIG welding: a –mode #1, b –mode #2, c –mode #3.

As seen on Figure 2, the fusion of the bottom layer is present, but the lack of arc oscillation limits the arc's ability to move across the joint, which can lead to inadequate fusion between the weld pool and the side walls. Without magnetically controlling arc, it remains concentrated in the center of the weld pool, leaving the side walls colder and at a higher risk of incomplete fusion (Akhonin et al. 2021). Magnetic induction plays a significant role in the narrow gap TIG welding of titanium alloys, particularly when magnetic arc oscillation is introduced. In the context of narrow gap TIG welding, where precise control of the weld pool and filler wire melting is crucial, magnetic induction helps optimize the process by influencing the thermal and fluid dynamics within the weld zone. When a magnetic field is applied during TIG welding, it interacts with the electrical current flowing through the arc, generating Lorentz forces that cause the arc to oscillate. This oscillation affects the heat distribution in the weld pool and directly influences the melting of the filler wire. As a result, the filler wire melts more efficiently and uniformly, which is particularly important in narrow gap welding where the filler material must penetrate deep into the joint while maintaining sidewall fusion.

One of the ways to increase the productivity of narrow gap TIG welding is to increase filler wire speed. To achieve the fill fusion with increased filler wire rate, it is required to increase arc oscillation amplitude and induction of the arc magnetic field. As seen from Figure 2, mode #3 with higher magnetic induction and higher filler wire speed, can introduce a number of challenges and potential issues that can negatively impact the welding process. One of the main risks of using high magnetic induction is that it can cause the arc to become overly unstable. At moderate levels, arc oscillation helps stabilize the arc and improve heat distribution. However, when the magnetic induction is too high, the arc can become excessively deflected or oscillate too vigorously, leading to erratic movements. This can result in poor control of the arc, making it difficult to maintain consistent heat input along the weld seam. An unstable arc can cause uneven melting of the filler wire and base material, increasing the risk of defects such as lack of fusion, porosity, or spattering. High magnetic induction can also cause uneven heat distribution in the weld pool. This uneven heat input can lead to incomplete fusion in narrow gap welding, where maintaining consistent heat throughout the joint is critical for ensuring full penetration and high-quality welds. As seen from Figure 2 and mode #2, the moderate magnetic induction values (standard values for narrow gap welding of titanium alloys), allows to obtain complete fusion of both bottom layer and side walls, that should result in defect less welded joint. Therefore, it is crucial to carefully control the magnetic induction within an optimal range to avoid these issues and achieve consistent, high-quality welds.

4 CONCLUSIONS

The model for narrow gap TIG welding with arc oscillation was developed for welding thick plates of titanium alloy Ti-6Al-4V with filler wire 2V. The finite element modelling was done on 3 modes – without arc oscillation, with arc oscillation and moderate filler wire speed and with arc oscillation and higher filler wire speed. The lack of arc oscillation limits the arc's ability to move across the joint, which can lead to inadequate fusion between the weld pool and the side walls. Magnetic induction, through arc oscillation, significantly improves the melting efficiency of the filler wire during narrow gap TIG welding of titanium alloys. It provides better control over heat distribution, enhances the stability of the arc, and ensures more uniform filler wire melting and deposition. While filler wire is necessary for building up the weld joint in narrow gap TIG welding of titanium, increasing its speed introduces several negative effects. It leads to an increasing of magnetic induction in the gap, which in turn, leads to an excessive arc instability, uneven heat distribution, weld pool turbulence, and negative impacts on filler wire melting. So, it is crucial to carefully control the magnetic induction within an optimal range to avoid these issues and achieve consistent, high-quality welds. Therefore, in a process that demands precise control over heat input and material deposition, high filler wire speeds can make it difficult to achieve the desired weld quality.

REFERENCES

Näsström, Jonas; Brueckner, Frank; Kaplan, Alexander FH. Imperfections in narrow gap multi-layer welding-potential causes and countermeasures. Optics and Lasers in Engineering, 2020, 129: 106011.

Akhonin, S. V., Belous, V. Y., Selin, R. V., & Kostin, V. A. Influence of TIG welding thermal cycle on temperature distribution and phase transformation in low-cost titanium alloy. In: IOP Conference Series: Earth and Environmental Science. IOP Publishing, 2021. p. 012012.

Paton, B. E.; Akhonin, S. V.; Prilutsky, V. P. Development of welding technologies in titanium component manufacturing. In: Ti 2011-Proceedings of the 12th World Conference on Titanium. 2012. p. 1585–1591.

Xu, W. H., Lin, S. B., Fan, C. L., & Yang, C. L. Prediction and optimization of weld bead geometry in oscillating arc narrow gap all-position GMA welding. The International Journal of Advanced Manufacturing Technology, 2015, 79: 183–196.

Fang, N., Xu, D., Huang, R., Qin, J., Jia, R., Cao, H., & Wu, P. Numerical simulation analysis of stress and strain of titanium alloy thick plate ultra-narrow gap welded joint by laser welding with filler wire. In: *Third International Conference on Optoelectronic Science and Materials (ICOSM 2021)*; SPIE, 2021. p. 456–465.

Pederson, Robert. *Microstructure and phase transformation of Ti-6Al-4V*. 2002. PhD Thesis. Luleå tekniska universitet.

Akhonin, S. V., Belous, V. Y., Selin, R. V., & Schwab, S. L. Effect of TIG-Welding on the Structure and Mechanical Properties of Low-Cost Titanium Alloy Ti-2.8 Al-5.1 Mo-4.9 Fe Welded Joints. In: *Materials Science Forum*. Trans Tech Publications Ltd, 2023. p. 105–110.

Akhonin, S. V., Belous, V. Y., Selin, R. V., & Berezos, V. A. Structure and properties of high-strength titanium alloy Ti-6.5 Al-3Mo-2.5 V-4Nb-1Cr-1Fe-2.5 Zr Welded Joints. Solid State Phenomena, 2021, 313: 82–93.

Welding and Related Technologies – Krivtsun et al. (Eds)

Charge and heat transfer from arc plasma to metal particle in GMA welding or PTA surfacing

I.V. Krivtsun
E.O. Paton Electric Welding Institute, Kyiv, Ukraine

A.I. Momot & D.V. Antoniv
E.O. Paton Electric Welding Institute, Kyiv, Ukraine
Taras Shevchenko National University of Kyiv, Ukraine

Binhao Qin
China-Ukraine Institute of Welding, Guangdong Academy of Sciences, Guangdong Provincial Key Laboratory of Advanced Welding Technology, Guangzhou, China

ABSTRACT: The charge and heat transfer from atmospheric pressure argon arc plasma to spherical metal particle is studied numerically. The particles of the radius 10^{-5}......10^{-4} m and 10^{-3} m are considered that corresponds to particles of additive material in plasma transferred arc (PTA) surfacing and droplets of electrode metal in gas metal arc (GMA) welding, respectively. The plasma is described in hydrodynamic approach with account for its thermal and ionization non-equilibrium near the particle. The hydrodynamic equations are formulated for quasi-neutral plasma, i.e. outside the sheath near the particle surface. In our study, the current density in plasma is $j_0 = 10^7$ A/m^2 that heats the argon plasma to about 15500 K. The electric current flowing through the particle in the arc plasma was found. The various components of the heat flux transferred to the particle in the arc plasma are found.

1 INTRODUCTION

Knowledge about charge and heat transfer from arc plasma to particles of additive material in PTA surfacing (Boulos et. al. 2023) and droplets of electrode metal in GMA welding (Lienert et al. 2011) is important for further development of these techniques. It is worth noting that most research on plasma - metal particle interactions consider currentless plasma, which corresponds to the plasma spraying conditions (Boulos et. al. 2023).

The major part of studies on the interaction between arc plasma and electrode metal droplets (size ~ 10^{-3} m) do not consider the changes of plasma parameters near the droplet, thermal and ionization non-equilibrium of plasma and the space charge sheath around metal droplet. As for fine particles with sizes in the range of 10^{-5}...10^{-4} m introduced into atmospheric pressure plasma arc during PTA surfacing, to the best of our knowledge, there are no studies in the current scientific and technical literature that explore the electrical interaction between current-carrying plasma and such particles.

The point of the present paper is to study numerically the electrical and thermal interactions between atmospheric pressure current-carrying argon arc plasma and a metal particle.

DOI: 10.1201/9781003518518-27

2 MODEL

The atmospheric pressure argon arc plasma with spherical metal particle is considered. Electrons and ions are adsorbed on the particle surface giving rise to the charge and heat transfer to the particle. Ions recombine on the surface and leave it as atoms with the temperature of the metal particle. Thus, the temperature T_h of heavy particles (atoms and ions) differs from the electron temperature T_e near the particle (plasma is in thermal non-equilibrium and, consequently, in ionization non-equilibrium). Our study focuses on the quasi-neutral plasma outside the space charge layer (sheath), which is formed near the particle surface. The characteristic time of plasma parameters establishing is much less than the temperature change of metal particle surface (Krivtsun et al. 2023). Hence, we consider the constant temperature of the particle surface $T_s = 1000$ K and study the stationary charge and heat fluxes to the particle. The evaporation of metal atoms and electron emission from the particle are negligible for the considered surface temperature. In our study, the current density in plasma is $j_0 = 10^{-3}$ A/m^2 that heats the argon plasma to about 15500 K. At this temperature, the presence of multiply charged argon ions can be neglected. The convective effects in the plasma are also ignored.

2.1 *Basic equations*

To describe the mass and heat dynamics in atmospheric pressure argon plasma we used the approach proposed in (Almeida et al. 2008). The continuity equations for electron ($\alpha = e$) and ion ($\alpha = i$) flux density $\mathbf{J}_\alpha = n_\alpha \mathbf{v}_\alpha$ have the form

$$\nabla \mathbf{J}_\alpha = \omega = k_i n_i n_a - k_r n_e^2 n_i. \tag{1}$$

Here n_α is the number density ($\alpha = a$ for atoms); $\mathbf{v}_\alpha$is the velocity of corresponding particles; k_i and $k_r$are the ionization and recombination rate constants, respectively. The production rate of plasma particles ω is due to collisions of electrons with atoms and the three-particle recombination involving an ion and two electrons.

Subtracting the equations (1), we obtain the continuity equation $\nabla \mathbf{j} = 0$ for current density $\mathbf{j} = e(\mathbf{J}_i - \mathbf{J}_e)$, where e is elementary charge. The mass-average plasma velocity is equal to zero $m_e \mathbf{J}_e + m_i \mathbf{J}_i + m_a \mathbf{J}_a = 0$ since plasma is at rest relative to the particle. Using that the electron mass is much less than the masses of ions and atoms, which are almost equal, we get $\mathbf{J}_a = -\mathbf{J}_i$.

The momentum equations for plasma particles can be found in (Zhdanov 2002)

$$-\nabla p_\alpha - n_\alpha Z_\alpha e \nabla \varphi + \sum\nolimits_\beta \mu_{\alpha\beta} n_\alpha n_\beta (\mathbf{v}_\beta - \mathbf{v}_\alpha) - \mathbf{R}_\alpha^T = 0. \tag{2}$$

Here $\alpha, \beta = e, i, a$; $p_\alpha = n_\alpha k T_\alpha$ is the partial pressure, where k is the Boltzmann constant; Z_α is the particle charge number; φ is the distribution of electric potential in plasma; $\nu_{\alpha\beta}$, $\mu_{\alpha\beta} = m_\alpha m_\beta / (m_\alpha + m_\beta)$ are the momentum transfer collision rate and reduced mass; $\mathbf{R}_\alpha^T = C_\alpha^{(e)} n_\alpha k \nabla T_e$ is the thermal diffusion force, where $C_\alpha^{(e)}$ are the thermal diffusion coefficients.

The sum of partial pressures is constant and equal to atmospheric pressure $n_i k T_e + n_i k T_h + n_a k T_h = p_0$. The last equation is used to find the atom number density distribution. It is possible to obtain from (2) the expressions for electron and ion fluxes and for current density

$$\begin{aligned}
\mathbf{J}_i &= \tfrac{1}{\Delta_{ei}}\left[(\zeta_i T_e - \zeta_e T_h) k \nabla n_i - (\zeta_e + \zeta_i) e n_i \nabla \varphi - \zeta_e k n_i \nabla T_h + (\zeta_i \tilde{C}_e^{(e)} - \zeta_e C_i^{(e)}) k n_i \nabla T_e\right] \\
\mathbf{J}_e &= -\tfrac{1}{\Delta_{ei}}\left[(\delta_i T_e - \delta_e T_h) k \nabla n_i - (\delta_e + \delta_i) e n_i \nabla \varphi - \delta_e k n_i \nabla T_h + (\delta_i \tilde{C}_e^{(e)} - \delta_e C_i^{(e)}) k n_i \nabla T_e\right] \\
\mathbf{j} &= \tfrac{e}{\Delta_{ei}}\left[(\gamma_i T_e - \gamma_e T_h) k \nabla n_i - (\gamma_e + \gamma_i) e n_i \nabla \varphi - \gamma_e k n_i \nabla T_h + (\gamma_i \tilde{C}_e^{(e)} - \gamma_e C_i^{(e)}) k n_i \nabla T_e\right]
\end{aligned} \tag{3}$$

where

$$\begin{aligned} \gamma_e &= \nu_{ea}\mu_{ea}(n_i + n_a), & \delta_e &= \nu_{ea}\mu_{ea}n_i - \nu_{ei}\mu_{ei}n_i, \\ \gamma_i &= \nu_{ia}\mu_{ia}(n_i + n_a), & \delta_i &= \nu_{ia}\mu_{ia}(n_i + n_a) + \nu_{ei}\mu_{ei}n_i, \\ \zeta_e &= \gamma_e - \delta_e = \nu_{ea}\mu_{ea}n_a + \nu_{ei}\mu_{ei}n_i, & \tilde{C}_e^{(e)} &= 1 + C_e^{(e)}, \\ \zeta_i &= \gamma_i - \delta_i = -\nu_{ei}\mu_{ei}n_i, & \Delta_{ei} &= \delta_i\gamma_e - \delta_e\gamma_i. \end{aligned} \tag{4}$$

The energy equations for electrons and heavy plasma particles have the form (Zhdanov 2002)

$$\nabla\left(\mathbf{q}_e + \frac{5}{2}kT_e\mathbf{J}_e\right) = e\mathbf{J}_e\nabla\varphi - \kappa_{eh}n_ik(T_e - T_h) - U_i\omega - w_{\text{rad}}, \tag{5}$$

$$\nabla\mathbf{q}_h = -e\mathbf{J}_i\nabla\varphi + \kappa_{eh}n_ik(T_e - T_h), \tag{6}$$

where $\mathbf{q}_e$ and $\mathbf{q}_h$ are the electron and heavy particles heat fluxes; κ_{eh} is the energy exchange frequency; U_i is the plasma forming gas (argon) ionization potential; w_{rad} denotes the energy losses due to radiation. The heat fluxes of electrons and heavy particles can be represented as

$$\mathbf{q}_e = -\lambda_e\nabla T_e + kT_en_e\left(A_i^{(e)}(\mathbf{v}_e - \mathbf{v}_i) + A_a^{(e)}(\mathbf{v}_e - \mathbf{v}_a)\right), \quad \mathbf{q}_h = -\lambda_h\nabla T_h, \tag{7}$$

where λ_e, λ_h are the thermal conductivity coefficients of electrons and heavy particles; $A_i^{(e)}$and $A_a^{(e)}$ are the kinetic coefficients.

Thus, we have four unknown distributions: $n_i(\mathbf{r})$, $\varphi(\mathbf{r})$, $T_h(\mathbf{r})$ and $T_e(\mathbf{r})$ that can be found from the system of differential equations: continuity equations for ion flux and current (3) and continuity equations for heat fluxes (5) and (6). The system should be supplemented with proper boundary conditions.

2.2 *Boundary conditions*

We consider plasma outside the sheath. Hence, the conditions are set on the sheath outer boundary S. The values of functions on this surface are denoted with subscript s, for instance n_{is}. The normal component of ion flux on the particle surface is $J_{is} = \mathbf{n}\mathbf{J}_i|_S = -(n_iv_B)|_S$,where $\mathbf{n}$is the outward unit normal vector to the surface and $v_B = (k(T_e + T_h)/m_i)^{1/2}$ is the Bohm velocity. The boundary condition for current density is

$$j_s = \mathbf{n}\mathbf{j}|_S = en_{is}\left[\frac{\bar{v}_{Te}}{4}\exp\left(-\frac{e\varphi_{sh}}{kT_e}\right) - v_B\right]_S, \quad \bar{v}_{Te} = \sqrt{\frac{8kT_e}{\pi m_e}}, \tag{8}$$

where $\bar{v}_{Te}$ is the mean thermal velocity of electrons; φ_{sh} is the plasma potential on the sheath outer boundary (potential of the particle is set equal to zero).

Temperature of heavy particles is equal to the temperature of the particle surface $T_h|_S = T_s$. The boundary condition for the electron temperature has the form $q_{es} = -0.5J_{es}kT_e$ (Almeida et al. 2008). We assume that unperturbed plasma (far away from the particle) is in thermal equilibrium $T_e = T_h = T_0$, i.e. $T_h|_{\mathbf{r}\to\infty} = T_0$. The unperturbed plasma is assumed also in the state of ionization equilibrium, thus the ratio between unperturbed ion n_{i0} and atom n_{a0} number densities is defined by the Saha equation.

The heat released in plasma as a result of electric current goes to thermal radiation $j_0^2/\sigma_0 = w_{\text{rad}}(T_0)$, where $\sigma_0 = e^2n_{i0}(\gamma_e + \gamma_i)/\Delta_{ei}$ is the conductivity of unperturbed plasma and $w_{\text{rad}}(T_0)$ is the power radiation loses per unit volume that can be described with the expression (58) from (Almeida et al. 2008). Computations were performed for atmospheric pressure argon plasma $p_0 = 1.013 \times 10^5$ Pa with current density $j_0 = 10^7$ A/m^2 that heats the plasma to $T_0 = 15522$K. Corresponding unperturbed ion number density is $n_{i0} = 1.931 \times 10^{23}$cm^{-3}.

2.3 *Heat flux*

The heat flux from plasma to the particle q_p consists of several components (Semenov et. al. 2016) $q_p = q_{p1} + q_{p2} + q_{p3} + q_{p4} + q_{p5} + q_{p6}$,where $q_{p1} = \lambda_e \nabla T_e|_S$ and $q_{p2} = \lambda_h \nabla T_h|_S$ are the heat fluxes due to electron and heavy particle thermal conductivity;

$$q_{p3} = -2.5kT_eJ_{es} - kT_eA_a^{(e)}(1 + n_i/n_a)J_{is} + kT_e\left(A_i^{(e)} + A_a^{(e)}\right)j_s/e \qquad (9)$$

is the heat flux due to convection; $q_{p4} = -U_iJ_{is}$ and $q_{p6} = (J_{is} - J_{es})A_f = j_sA_f/e$ are the heat fluxes due to ion recombination and electron absorption on the particle surface, where A_f is the work function of the particle metal. In our study we have used the value for steel $A_f = 4.5$eV.

The potential drop between sheath outer boundary and the particle surface gives to the ion the additional kinetic energy $e\varphi_{sh}$, which is released in inelastic collision with the particle surface. Corresponding heat flux is $q_{p5} = -e\varphi_{sh}J_{is}$.

3 RESULTS AND DISCUSSION

The considered problem has a cylindrical symmetry; thus, it is two-dimensional in the cylindrical coordinate system placed in the center of the particle and z-axis directed along $\mathbf{j}_0$. Unknown functions depend on r and z variables (see Figure 1). The system of partial differential equations was solved numerically with finite element method using the FlexPDE program (ver.7.22). The kinetic coefficients from the equations can be found in the appendix of (Krivtsun et al. 2023).

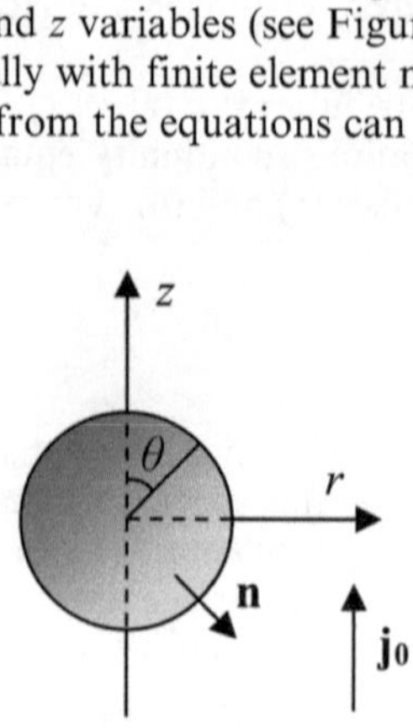

Figure 1. Mutual displacement of particle and coordinate system.

We performed calculations for two cases, the first one with approximation that electron temperature is constant. This is justified by our previous calculations (Krivtsun et al. 2023) for argon plasma without current, which showed a decrease in electron temperature near the particle surface of up to 15% for particles with radius $a = 10^{-4}$ m. In this case, the last terms in (3), which contain∇T_e, are equal to zero and we don't need continuity equation for electron heat flux (5). The second case is the general one $T_e \neq const$.

The present calculations for temperature distributions show that the electron temperature decreases up to 30% near the particle with radius $a = 10^{-3}$ m (see blue solid line in Figure 2). Moreover, T_e depends on the angle θ from the direction of current (z-axis) as consequence the electron temperature is different in points $z = a$ and $z = -a$ in Figure 2. The values of mean electron temperature on the particle surface $\bar{T}_{es}$ are presented in Table 1. According to the boundary condition $T_h|_S = T_s$, where $T_s = 1000$ K in our calculations. On the distance about particle radius from the surface plasma becomes isothermal. Further temperature reaches its unperturbed value T_0 on the distance ~ $20a$ from the particle surface.

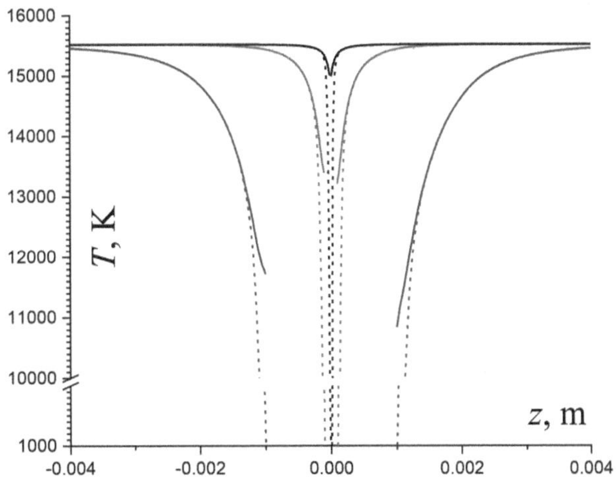

Figure 2. Electron temperature T_e (solid lines) and heavy particle temperature T_h (deshed lines) distributions along z-axis ($r = 0$) for $j_0 = 10^7$ A/m^2 and particle radii $a = 10^{-5}$, 10^{-4} and 10^{-3} m.

Table 1. Mean electron temperature on the particle surface $\bar{T}_{es}$(K), ion surface number density n_{is}, ion surface flux density J_{is} and current through the particle I_s for $j_0 = 10^7$ A/m^2 and particle radii $a = 10^{-5}$, 10^{-4} and 10^{-3} m.

a (m)	$\bar{T}_{es}$(kK)	n_{is}(m^{-3})		J_{is}(m^{-2}s^{-1})		I_s(A)	
	$T_e \neq const$	$T_e = const$	$T_e \neq const$	$T_e = const$	$T_e \neq const$	$T_e = const$	$T_e \neq const$
10^{-5}	1.50	6.24×10^{22}	6.12×10^{22}	-1.16×10^{26}	-1.12×10^{26}	8.83×10^{-5}	7.64×10^{-5}
10^{-4}	1.33	1.42×10^{22}	9.19×10^{21}	-2.64×10^{25}	-1.59×10^{25}	1.98×10^{-2}	1.17×10^{-2}
10^{-3}	1.18	8.57×10^{21}	1.74×10^{21}	-1.59×10^{25}	-2.84×10^{24}	9.42	2.26

Figure 3 shows that the decrease of T_h near the particle surface leads to a decrease of plasma particle density and an increase of neutral particle density, since the pressure is constant. This figure also illustrates the range where the particle perturbs plasma density.

The decrease of n_i is more considerable near the bigger particle. The averaged value of plasma density on the surface n_{is} is presented in Table 1. One can see that approximation $T_e = const$ overestimates values of n_{is} as well as values of ion flux on the particle surface $J_{is} = -n_{is}v_B$.

One can see in Figure 4 that current flows into the particle ($j_s < 0$) from one side and flows out ($j_s > 0$) from the other side. The highest value of j_s is reached for $\theta = 0$, it is much less than the current density in unperturbed plasma ($j_s/j_0 \approx 0.02\ldots0.1$). Current density distribution become considerably asymmetric for $a = 10^{-3}$. The electric current flowing through the particle

$$I_s = 2\pi a^2 \int_{j_s>0} j_s \sin\theta d\theta \qquad (10)$$

is presented in Table 1

Further we discuss the heat flux on the particle. The heat flux density components, which depend on the angle θ, were averaged over the particle surface and denoted with the bar $\bar{q}_{pi}$.

Table 2 presents the results of calculations for $T_e = const$ (q_{p1} and $\bar{q}_{p6}$ are not listed in the table since they are equal to zero). One can see that the heat flux components are arranged as $q_{p4} > q_{p2} > \bar{q}_{p5} > \bar{q}_{p3}$. Thus, the main mechanisms of heat transfer from plasma to the particle are the thermal conductivity of heavy particles and heat flux due to ion recombination. This

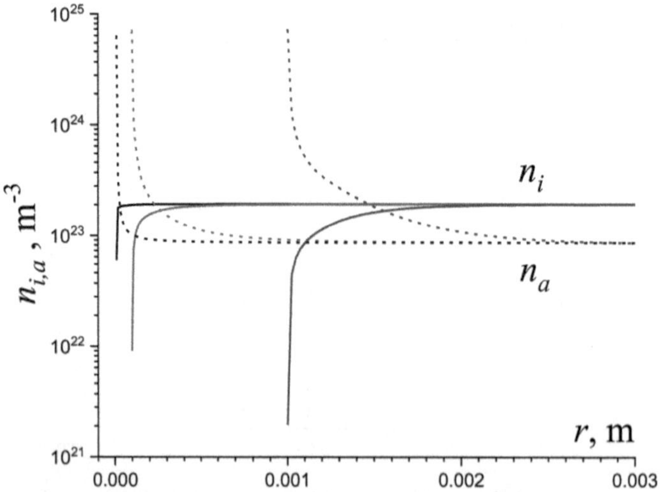

Figure 3. Ion and atom number density distributions along r-axis ($z = 0$) for $j_0 = 10^7$ A/m^2 and particle radii $a = 10^{-5}$, 10^{-4} and 10^{-3} m, $T_e \neq const$.

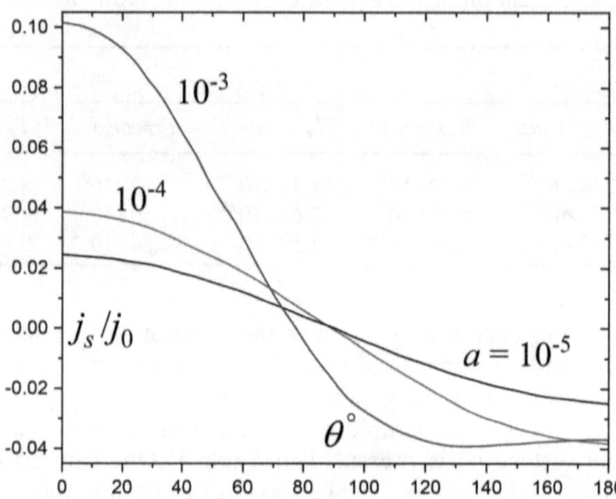

Figure 4. Normalized current density (8) on the particle surface $j_s j_0$ vs. θ for $j_0 = 10^7$ A/m^2 and particle radii $a = 10^{-5}$(black), 10^{-4} (red) and 10^{-3} m (blue), $T_e \neq const$.

conclusion is in agreement with those in (Krivtsun et al. 2023). According to the boundary condition for electron temperature $q_{es} = -0.5 J_{es} k T_e$ and Equations (7) and (9) the sum of $q_{p1} + q_{p3}$ should be equal to $-2kT_e J_{es}$.This equality is not satisfied for $T_e = const$. The input of $\bar{q}_{p3}$ in total heat flux is the least, thus the discrepancy is not considerable.

Table 2. Heat flux from plasma to the particle $\bar{q}_p$ and its components in W/m^2 and heat power $Q_p = 4\pi a^2 \bar{q}_p$ in W for $j_0 = 10^7$ A/m^2 and particle radii $a = 10^{-5}$, 10^{-4} and 10^{-3} m, $T_e = const$.

a (m)	q_{p2}	$\bar{q}_{p3}$	$-2kT_e\bar{J}_{es}$	q_{p4}	$\bar{q}_{p5}$	$\bar{q}_p$	Q_p	Q_p/V
10^{-5}	2.16×10^8	4.35×10^7	4.95×10^7	2.92×10^8	1.15×10^8	6.66×10^8	0.837	2.00×10^{14}
10^{-4}	5.72×10^7	6.73×10^6	1.13×10^7	6.66×10^7	2.63×10^7	1.57×10^8	19.7	4.71×10^{12}
10^{-3}	3.57×10^7	3.67×10^6	6.81×10^6	4.01×10^7	1.22×10^7	9.16×10^7	1151	2.75×10^{11}

Table 3 shows that in the general case $T_e \neq const$ the components q_{p2} and q_{p4} of the heat flux to the particle give the main contribution to the total heat flux q_p that is less than in the case $T_e = const$. The last approximation is valid for small particles ($a = 10^{-5}$) and the difference in the heat flux increases with the particle radius. This is explained by the fact that the electron temperature and ion flux decrease considerably near larger particles (see Tab. 1 and Figure 2).

Table 3. Heat flux from plasma to the particle $\bar{q}_p$ and its components in W/m^2 and heat power $Q_p = 4\pi a^2 \bar{q}_p$ in W for $j_0 = 10^7$ A/m^2 and particle radii $a = 10^{-5}$, 10^{-4} and 10^{-3} m,$T_e \neq const$.

a (m)	$\bar{q}_{p1}$	$\bar{q}_{p2}$	$\bar{q}_{p3}$	$\bar{q}_{p4}$	$\bar{q}_{p5}$	$\bar{q}_p$	Q_p	Q_p/V
10^{-5}	5.91×10^6	2.62×10^8	4.04×10^7	2.82×10^8	1.08×10^8	6.99×10^8	0.877	2.09×10^{14}
10^{-4}	2.65×10^6	5.33×10^7	3.17×10^6	4.00×10^7	1.36×10^7	1.13×10^8	14.2	3.38×10^{12}
10^{-3}	4.92×10^5	1.76×10^7	4.13×10^5	7.17×10^6	2.39×10^6	2.81×10^7	353	8.42×10^{10}

The heat power per volume of the particle Q_p/V ($V = 4\pi a^3/3$) considerably decreases with the radius growth (see last columns of Tab. 2 and 3). It means that a small particle is heated and melted faster than a large particle since the heat capacity and heat of fusion are proportional to the particle volume.

ACKNOWLEDGMENTS

This work was supported by Guangdong Academy of Sciences Project of Science and Technology Development (2020GDASYL-20200301001), The National Key Research and Development Program of China (grant number 2020YFE0205300), and China Scientific Research Fund Project (G2023030020L).

REFERENCES

Almeida, N. A., Benilov, M. S., & Naidis, G. V. 2008. Unified modelling of near-cathode plasma layers in high-pressure arc discharges. *Journal of Physics D: Applied Physics*, 41(24): 245201.

Boulos, M. I., Fauchais, P. L., & Pfender, E. (Eds.) 2023. *Handbook of thermal plasmas*. Springer Nature.

Krivtsun, I. V., Momot, A. I., Antoniv, D. V., & Qin, B. 2023. Characteristics of atmospheric pressure Ar-plasma around a spherical particle: Numerical study. *Physics of Plasmas*, 30(4): 043513.

Lienert, T., Siewert, T., Babu, S., & Acoff, V. (Eds.). 2011. *Welding fundamentals and processes*. ASM International.

Semenov, I. L., Krivtsun, I. V., & Reisgen, U. 2016. Numerical study of the anode boundary layer in atmospheric pressure arc discharges. *Journal of Physics D: Applied Physics*, 49: 105204.

Zhdanov, V. M. 2002 *Transport processes in multicomponent plasma*. CRC Press

Welding and Related Technologies – Krivtsun et al. (Eds)

Modeling of temperature fields distribution and stress-strain state of Mg-Al alloy FSW joints

M.A. Khokhlov, V.A. Kostin, J.A. Khokhlova, A.G. Poklaytskiy & O.O. Makhnenko
E.O. Paton Electric Welding Institute, Kyiv, Ukraine

ABSTRACT: Experimental equipment for friction stir welding (FSW) allows to obtain thin high-quality butt joints on optimal welding modes at different linear speeds. Mathematical modeling allows to analyze the stress-strain state and thermomechanical processes in the FSW joint zone, and to predict the operational properties, strength and service life of welded structures. The temperature distributions from the volume source of heating during FSW were visualized, and the residual deformations and stresses in the zone of butt-welded joints of thin sheets of magnesium alloys were numerically determined using finite-element models. The width of the HAZ decreases, maximum residual stresses increase by 10. . .15% while residual plastic deformations decrease with an increase in the linear speed of welding. For future, it is advisable to determine the balance between linear and rotational speed of the FSW tool to obtain greater uniformity of the structure and to reduce heat transfer to the metal during welding.

1 INTRODUCTION

The development of effective technical and technological solutions related to increasing the productivity, controllability, and economy of constructions of the near future inspire engineers to use magnesium alloys (Yanga 2023). Regarding the welding of magnesium alloys, we have an experience of using an ecological and energy-saving technology – friction stir welding (FSW), which is used to connect parts with a relatively simple geometry in the solid phase (Figure 1). The experience (Krasnovsky 2019) of using a FSW tool in the form of a smooth pin with a concave shoulder is statistically successful, because it avoids deformation when joining light and ductile alloys, welds are formed without voids and contain locally strengthened zones along the lower contour of the weld core. The idea of the following experiments was the mathematical modeling of thermal and deformation processes during the FSW using fundamental regularities (Tsaryk 2022). Such a comprehensive approach to assessing the quality of the formation of a welded joint is being developed for further use in statistical scaled models to predict the operational properties, strength and service life of welded structures made of thin magnesium alloys. The purpose of this study is to obtain high-quality FSW butt joints of magnesium alloy at a constant tool rotational speed and at different linear welding speeds, to determine by indentation the mechanical parameters of the material in the FSW zone, to experimentally measure surface temperature fields during FSW and to determine temperature fields, distributions of deformations and residual stresses in FSW butt joints by methods of mathematical modeling.

2 MATERIALS AND METHODS

Welding of butt joints of thin sheets magnesium alloy of the Mg-Al-Zn system (Table 1) with a thickness of 2 mm was carried out on experimental laboratory welding equipment for FSW at a constant tool rotational speed of 1420 rpm and at different linear speeds (8 m/h, 16 m/h, 24 m/h). A comparative assessment of temperature fields (°C) during the FSW experimentally

DOI: 10.1201/9781003518518-28

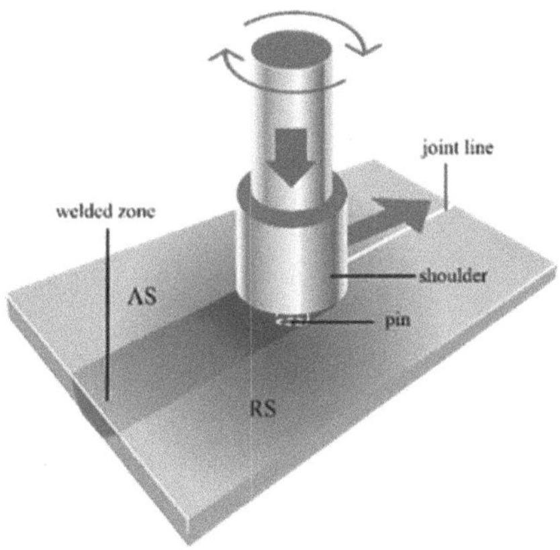

Figure 1. The scheme of FSW process.

was done with a Fluke Ti25 infrared thermal imager, and the kinetics of temperature field changes during the FSW were simulated in COMSOL Multiphysics Thin sections of welded joints were prepared and treated with an aqueous solution of a mixture of 5% hydrochloric acid and 5% orthophosphoric acid to determine the features of the macro- and microstructure by optical microscopy and physical and mechanical properties by indentation. Indentation at a load of 100 g with automatic determination of the modulus of elasticity (E) was carried out with a three-sided Berkovich indenter on a computerized device "Micron-Gamma" according to standard methods and ISO/FDIS 14577–1:2015; Metallic materials – Indentation test for hardness and materials parameters. Experimental determination of Young's modulus was carried out to clarify the data used for welding simulation. Mathematical modeling of the FSW process was carried out on magnesium alloy plates with a thickness of 2 mm at a constant tool rotational speed of 1420 rpm and for different linear welding speeds of 8 m/h, 16 m/h, 24 m/h using a specialized finite element analysis program of our own development.

3 RESULTS AND DISCUSSION

Determination of the modulus of elasticity (E) was carried out for 5 types of FSW weld textures – base metal, TMAZ, top, bottom and center of the core (Figure 2). In the central part of the core E=30 GPa, the zone contains an ellipsoidal structure and textures drawn from pressing and rotating the tool in the upper and lower parts of the weld. For the top part of the weld E=36 GPa. At the bottom of the core at 50-150 μm along the lower edge of the sample, the highest value of E=90 GPa was determined. That is, an increase in elasticity by 2–3 times compared to the base metal E= 35 GPa. For the TMAZ zone, the modulus of elasticity E=30 GPa. A comparison of the shape and depth of the indentation diagrams shows a more plastic state of the base metal, and in the core zone below the middle of cross-section there is an increased elastic response of the material surface. The depth of the indenter dipping into the base material is greater and is 7.57 μm, and in the core zone below the middle of the cross-section it is 5.44 μm. Since such a significant heterogeneity in textures of zones in the FSW joints creates a high level of elastic-deformed state in the thin metal, it ensures the traditional statistics of the fracture by the TMAZ zone in the tensile strength tests, therefore, an additional heat treatment is usually recommended, after which the condition can be normalized by about 10 %.

According to the experimental thermogram of the welding process recorded by the thermal imager, it was approximately determined that the outer surface of the magnesium alloy and the tool during welding has a temperature of no more than 355.8°C. The model of the temperature fields with the specified welding parameters (Figure 3) shows a similar temperature at a small distance from the point of contact of the cylindrical FSW tool with the metal.

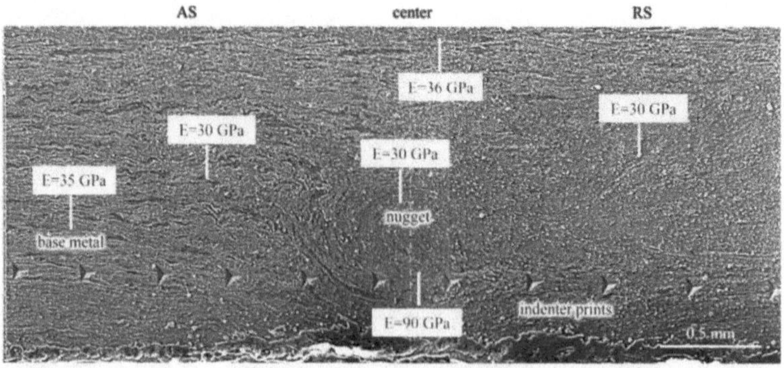

Figure 2. Microstructure of the central part of the FSW weld.

After passing of the FSW tool, the temperature is at the level of 410°C, and in the zone of the weld core, where a crushed structure is formed with an increase in elasticity by 2-3 times, the temperature is 500°C. Thermal cycles during FSW and the distribution of maximum temperatures along the volume of the plates depending on the welding speed were plotted by the results of the finite element analysis. The maximum temperature of the thermal cycle is approximately 600°C, and when the linear welding speed is increased above 16 m/h, it changes to 450°C. That is, the maximum temperature in the contact zone of the tool with the material of the plates is mainly determined by the rotational speed and by the pressing force of the tool against the plates. The width of the temperature fields in the joint zone (Figure 4) is 8 mm, 6 mm, and 4 mm, with increasing linear speeds of 8 m/h, 16 m/h, and 24 m/h respectively.

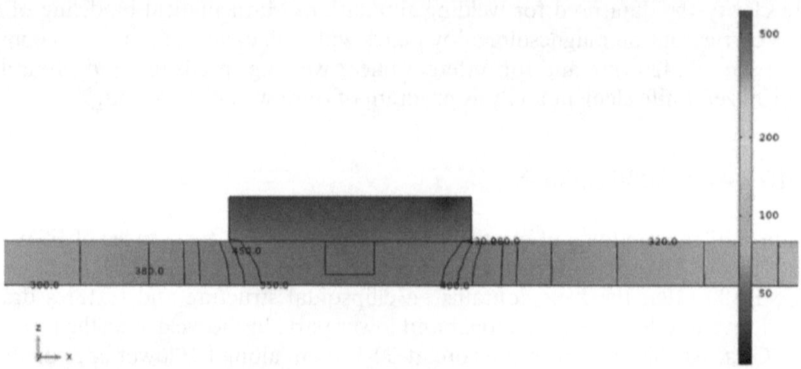

Figure 3. Model of the temperature fields of the FSW process at a linear speed of 8 m/h (movement of the tool along the x axis).

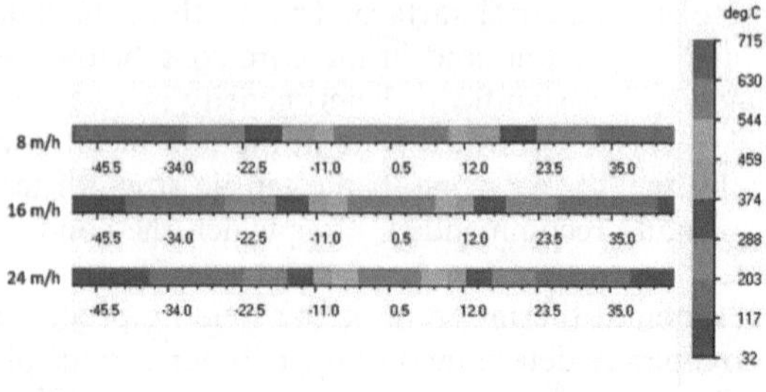

Figure 4. Distribution of temperature fields (mm) during FSW of 2 mm-thick magnesium alloy plates at different linear welding speeds.

Calculations of the maximum residual tensile stresses showed that they reach the yield point of the material of 140 MPa (Figure 5a). The level of the maximum calculated residual plastic shrinkage deformations is relatively low, up to −0.35% (Figure 5b). When the linear speed of welding is in-creased, the maximum residual tensile stresses increase by 10...15%, which is associated with an increase in the temperature gradient during welding. It was also established that residual plastic deformations (longitudinal) have a slight decrease in level with an increase in the linear speed of welding.

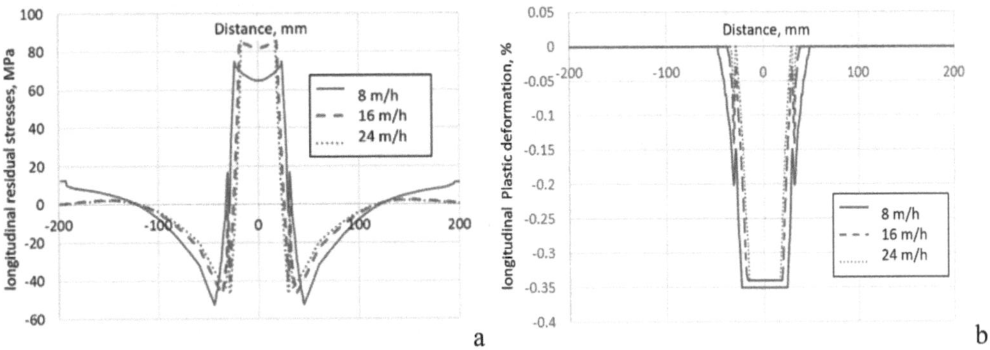

Figure 5. Distribution of longitudinal residual stresses (a) and plastic deformations (b) in the butt joint of 2-mm-thick magnesium alloy plates at different FSW linear speeds.

4 CONCLUSIONS

High-quality welded joints of thin sheets of magnesium alloy are obtained at a constant rotation speed of the FSW tool of 1420 rpm and linear welding speeds of 8 m/h, 16 m/h, 24 m/h. A clear textured border is formed between the weld core and TMAZ on the retreating side, the weld core contains a refined ellipsoidal structure with 2-3 times increase in the modulus of elasticity compared to the base metal. The calculated temperature field characterizing the FSW process correlates with the experimentally determined surface temperature and has a maximum temperature of 600°C in the core zone, so the metal in the joint zone is in a plastic state. Modeling of the kinetics of changes in temperature fields during the FSW process showed that the width of the HAZ decreases with an increase in linear speed. It was established that with an increase in the linear speed of welding, the maximum residual stresses increase by 10...15%. Residual plastic deformations decrease with an increase in the linear speed of welding. To obtain uniform structure of the FSW weld of thin sheet magnesium alloy, it can be effective to reduce the heat transfer to the metal by increasing the FSW linear speed, or by reducing the rotational speed of the FSW tool, i.e., in the future, it is advisable to determine the balance of the linear speed and the rotational speed of the FSW tool.

REFERENCES

Krasnovsky, K. 2019. Influence of tool shape for friction stir welding on physicomechanical properties of zones of welds of aluminium alloy EN AW 6082-T6. The Paton Welding Journal, 7. 7–11.

Tsaryk, B.R. 2022. Mathematical model of determination of residual stresses and strains in friction stir welding of aluminium alloy. The Paton Welding Journal, 9. 37–44.

Yanga, Y. 2023. Review. Research advances of magnesium and magnesium alloys worldwide in 2022. Journal of Magnesium and Alloys 11(8), 2611–2654.

Session 5. Materials and technologies for medicine and ecology of welding production

Welding and Related Technologies – Krivtsun et al. (Eds)
© 2025 The Author(s), ISBN 978-1-032-85176-1
Open Access: www.taylorfrancis.com, CC BY-NC-ND 4.0 license

Current state and prospects for development in welding of biological tissues

S.V. Tkachenko, O.V. Romanenko & N.A. Chvertko
Paton Welding Institute, The National Academy of Sciences of Ukraine, Kyiv, Ukraine

A.G. Dubko
Paton Welding Institute, The National Academy of Sciences of Ukraine, Kyiv, Ukraine
National Technical University Kiev Polytechnic Institute Igor Sikorsky, Kyiv, Ukraine

ABSTRACT: The results of the research of the Paton Welding Institute in the field of high-frequency welding of biological tissues for their connection, coagulation and cutting are presented. The authors share their experience in creating specialized equipment, tools and algorithms. In laboratory and clinical conditions, complex studies of the behavior of various types of various tissues under the influence of high-frequency current were carried out, new original process algorithms were proposed and tested, which ensure high quality of the obtained connections and shorten the recovery time of operated tissues.

The use of developed technologies and equipment in practical surgery ensures: bloodless, fast, low-traumatic for the patient surgical interventions, reliable hemostasis without the use of suture materials; reduction of blood loss by more than 50%; reducing the duration of operations by 20…50%; rapid rehabilitation of patients after surgery.

Prospects for further development of technologies for high-frequency welding are considered.

1 INTRODUCTION

There are a large number of companies producing medical coagulators. This is reviewed by Marinsky et al. (2023). There should be mentioned such manufacturers as Medtronic, Bovie/Aaron and Ethicon (USA), KLS Martin, Karl Storz and Bowa Medical (Germany), etc. These companies present in the market and invest a lot of money into RnD and its production.

Most electrosurgical devices operate in the cutting and coagulation modes and provide operation with bipolar and monopolar connections. The operating frequency is between 350 kHz - 1 MHz. The output power (10 W – 400 W) is adjusted depending on the area where the device is used. The coagulators usually have up to a dozen different modes. To display information, LEDs, digital indicators or LCD touch panels are used. Some devices can be used in conjunction with other equipment - as an argon plasma coagulator, ultrasound surgery devices, and can be connected to the network. Also, manufacturers try to simplify the procedure for setting up and changing modes: easy switching between previously set parameters of modes; algorithms of self-testing, self-adjustment and adaptation for every case are being developed.

Since 1993, the Paton Welding Institute (PWI) together with surgeons of various specialties, have been engaged in conducting experiments, developing the technologies, equipment and electrosurgical tools for connecting, processing and cutting a living biological tissues. The idea of the possibility of connecting biological tissues with the help of electric welding technologies belongs to Academician B.E. Paton. It was tested for the first time with a positive result by specialists of the National Institute of Surgery and Transplantology A.A. Shalimov of the Academy of Medical Sciences of Ukraine and “Okhmatdyt” Hospital - Paton & Ivanova (2009).

DOI: 10.1201/9781003518518-29

2 TISSUE WELDING – HISTORY AND RECENT RESULTS

When welding biological tissues, complex processes of converting electrical and mechanical energy into thermal energy take place. This type of welding can be attributed to contact resistance welding - Lebedev & Dubko (2012).

Scientific developments on the implementation of technologies, equipment and tools for welding of living tissues into medical practice have been tested at scientific international conferences, presented in rated scientific publications and protected by a number of patents –Tkachenko et al. (2022), Paton et al. (2006).

Over 30 years, PWI specialists have created several generations of equipment and tools for welding of living tissues in medicine and veterinary medicine. In 2011, a new generation high-frequency electrosurgical device EKVZ-300 TM “PATONMED” was developed (Figure 1).

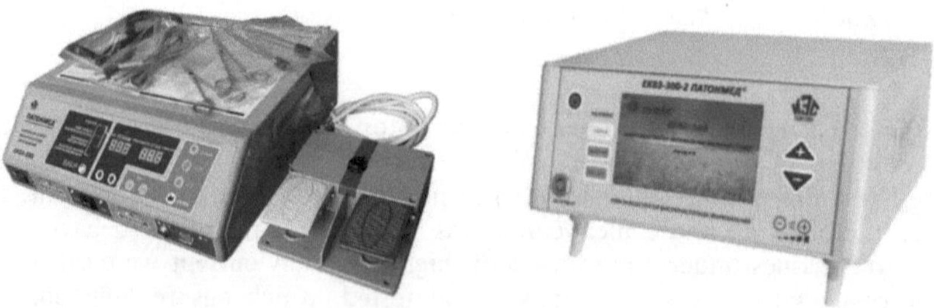

a b

Figure 1. RF generators EKVZ-300 “PATONMED” (a) and EKVZ-300-2 with LCD screen.

Operating frequency of EKVZ-300 “PATONMED” device - 440 kHz, maximum output power 300 W, weight 7.5 kg. The device implements a new method of bipolar high-frequency welding of living tissues, which allows, to expand the possibilities of surgical interventions, minimize the thermal effect on the operated tissue and lead to obtaining high-quality welded joints with the restoration of functions operated living tissues, which is confirmed by histological studies. A registration certificate and permission for serial production of the device EKVZ-300 “PATONMED” have been received. This device is included in the State Register of Medical Equipment and Medical Products of Ukraine.

Nowadays, welding of living biological tissues is widely used in the following areas of surgical practice: general abdominal surgery; thoracic surgery; otorhinolaryngological operations; vascular surgery; gynecology; proctology; urology; traumatology The PWI ABI-3500 device was developed for irreversible electroporation in the treatment of oncological diseases. The development of this direction is somewhat delayed due to the need to develop an appropriate tools.

Leading surgeons have confirmed that the method of high-frequency live tissue welding, in comparison with traditional surgical methods, provides: bloodless, fast, convenient for the surgeon and minimally traumatic for the patient performance of surgical interventions, reliable hemostasis; reliability and tightness of the connection of tissue; absence of suppuration and tissue necrosis; reduction of blood loss by more than 50%; reducing the duration of operations by 20…50%; high ablasticity of operations; quick postoperative rehabilitation.

Various electrosurgical instruments are shown in the Figure 2.

In 2023-2024, numerous experiments were conducted to develop or improve the algorithms for live tissues welding, the goal of which was to obtain reliable energy-saving technologies. The strength of the connected tissues was measured in laboratory conditions.

We present some results of experiments on welding of live biological tissues of various types.

Welding of an artery with a bipolar clamp is shown in Figure 3a - Lebedev & Dubko (2020). The average strength of the connection of arteries with a diameter of 2-8 mm was seven times greater than the normal systolic pressure (940 ± 84 mm Hg). Figure 3b shows a test of the hydraulic pressure force of 1269 mm Hg endured by welded carotid artery of a pig.

The PWI has developed a prototype tools for end-to-end welding of the intestines by open and laparoscopic methods. Unlike traditional staplers, no metal staples are left in the body.

a b c d e

Figure 2. Electrosurgical instruments: a) tweezers; b) clamps; c) for laparoscopic operations; d) to attach the retina to the choroid of the eye; e) for intestine anastomosis operation.

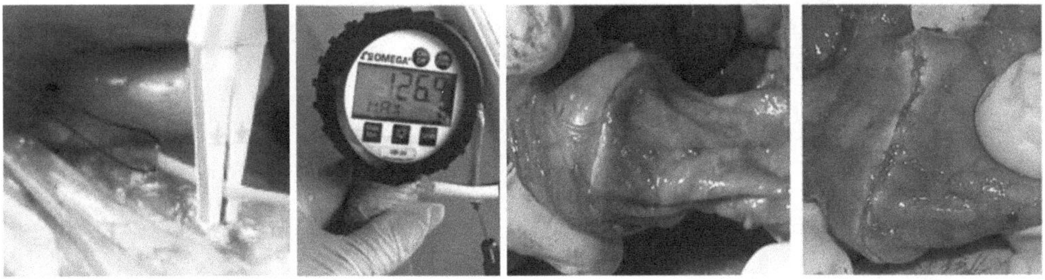

a b c d

Figure 3. a - welding of an artery, b – measuring the hydraulic pressure of the carotid artery, c, d- appearance of colon anastomosis.

The appearance of electro welding intestinal anastomosis with a minimal zone of thermal influence is shown in Figure 3c, d. The high quality of the connection is confirmed by histological studies.

The latest results are - the hydraulic strength of welded anastomosis was equal to 163-207 mm Hg, which was 6 - 8 more than the strength of the anastomosis obtained using a stapler.

A vivid example of the advantages of welding in comparison with laser coagulation is the welding of the retina of the eye to its choroid (Figure 4a). This method is less traumatic, quick and allows to reduce postoperative complications (Umanets et al. 2014). To achieve the positive results, the hardware was changed and the special software algorithms were developed.

In 2023-2024, research was carried out on the welding of tendons removed from a pig (Figure 4b). PWI scientists obtained high strength results of the welded joint of tendons (thickness 2.5 mm, width 3.5 mm) - in the range from 1.5 to 2.1 kg. Further research and reliable results in this direction can provide a revolutionary technique for surgeons!

Among other interesting studies performed at the PWI - lung resection; connection of the epineurium of the segments of the severed sciatic nerve by the method of multi-point electric welding (Molotkovets 2019); high-frequency bipolar electric welding for injuries of the dura mater (Zabolotnyi & Kvasha. 2023).

In the presence of developed reliable equipment and tools, it is especially important to create effective welding algorithms that allow obtaining a strong connection without tissue necrosis. The algorithms are based on the feedback got by measuring the electric parameters of tissue. It is a serious scientific research and even a kind of art to combine all the parameters – power, frequency, PWM duty cycle, and longitude for each stage of the process to receive the high-strength weld connection of live tissues.

Further researches of technologies and equipment for high-frequency welding relate to the development of hybrid welding technologies using high-frequency electromagnetic, ultrasonic, and thermal energies and the development of specialized equipment and tools for performing surgical interventions. We are also developing the portable equipment, electrosurgical technologies for welding living tissues in military field conditions, which will significantly reduce the time of operations, reduce blood loss, and reduce the inflammation and complications.

Department No.17 of the PWI conducts extensive international cooperation. We have developed equipment for clients from Canada and USA, we are undergoing FDA certification. Many samples have been sold to Chinese companies; with one of them we are negotiating the sale of a license for serial production. A cooperation agreement was signed with a company from Brazil.

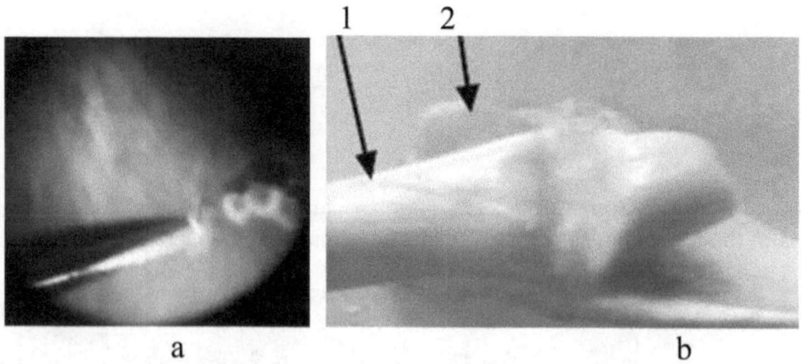

Figure 4. a - welding of the retina to the choroid of the eye with a monopolar instrument; b - welding connection of tendons 1, 2.

3 CONCLUSIONS

1. Analysis of the technical characteristics of modern medical electrocoagulators shows high competition in the market of electrosurgical equipment and instruments.
2. The high-frequency welding electrocoagulators of the EKVZ-300 series, developed at the PWI, meet the world standards.
3. The developed equipment, technologies and algorithms allow the practical surgeons to replace traditional surgical methods, in many areas of medicine.
4. Prospects for further development of technologies and equipment for high-frequency welding in medicine and veterinary medicine are considered.
5. The PWI engineers develop new models of electrosurgical devices by the contracts; create specific modifications of existing equipment and tools.

REFERENCES

Lebedev, A.V., Dubko A.G. 2012. Peculiarities of applying the theory of contact welding of metals to the welding of living tissues. *Technical electrodynamics*. №2: 187–192.

Lebedev, A.V., Dubko, A.G. 2020. Use of Electric Welding of Living Tissues in Surgery. *Biomedical Engineering*, №1: 1–6.

Marinsky, G.S., Tkachenko, V.A., Bysko, V.O., Podpryatov, S.E., Podpryatov, S.S., Grabovsky, S.D., Tkachenko, S.V. 2023. High-frequency equipment for live tissue welding (review). *The Paton Welding Journal*, № 1: 23–30.

Molotkovets, V.Yu. 2019. Morphological changes of the nerve after restorative treatment using electric welding technology (Experimental study). *Endovascular neuroradiological surgery*, No. 1(27): 91–98.

Paton, B.E., Ivanova, O.N. 2009. Tissue-saving high-frequency electric welding surgery. *Atlas. Kyiv. Naukova Dumka*: 197 p.

Paton, B.E., Lebedev, V.K., Lebedev, A.V. and others. 2006. Patent of Ukraine No. 77064. The method of welding biological tissue and the device for welding biological tissue (variants). Publ. 15.10.2006. Bull. No.10/2006.

Tkachenko, V.A., Marinsky, G.S., Podpryatov, S.E., Chvertko, N.A., Tkachenko, S.V., Chernets, O.V., Grabovsky, S.D. 2022. Patent of Ukraine No. 151770 for a utility model. Electrocoagulator high-frequency welding EKVZ - 300 "Patonmed". Publ. 09. 14, 2022, Bull. No. 37/2022.

Umanets, N.N., Pasechnikova, N.V., Dumbrova, N.E., Molchanyuk, N.I. 2014. Ultrastructural changes in the retina and choroid of a rabbit in the late stages after exposure to various modes of high-frequency electric welding of biological tissues. *Ophthalmology Eastern Europe*, No. 2 (21): 98–107.

Zabolotnyi D, Kvasha O. 2023. Experimental development of the method of seamless connection of the dura mater using high-frequency bipolar electric welding and substantiation of its greater tightness, compared to the traditional method of connection using suture material. *International scientific journal «Grail of Science»*, № 31. P. 447–458.

Session 6. Non-destructive testing and technical diagnostics

Welding and Related Technologies – Krivtsun et al. (Eds)

Determination of a TV tower verticality using UAVs, RTK and photogrammetry

L.M. Lobanov, D. Stelmakh, I. Shkurat & V. Savitsky
E.O. Paton Electric Welding Institute of the National Academy of Sciences of Ukraine, Kyiv, Ukraine

R. Toldo & A. Alessi
3Dflow SRL, Verona, Italy

ABSTRACT: The study of a TV tower using unmanned aerial vehicles equipped with an RTK module and the photogrammetry method demonstrated effectiveness in determining the verticality of the infrastructure object. The RTK technology was used to determine the geospatial position of the digital camera, allowing the creation of detailed 3D models. A methodology was proposed for determining the deviation of the central axis of the TV tower from the vertical direction based on data obtained through photogrammetry.

1 INTRODUCTION

Large structures, such as television, cellular, and power towers, play an important role in providing communication services and energy transmission. However, these structures are constantly subjected to natural forces such as wind, rain, snow, temperature fluctuations, and seismic activity. Over time, these factors affect the structural materials, leading to wear and a gradual loss of strength. Furthermore, inadequate maintenance of these structures can accelerate deterioration, increasing the risk of accidents and failure of critical infrastructure elements.

Diagnosing the verticality of large structures is crucial for their maintenance and safe operation. Deviations from verticality can indicate deformation, substrate settlement, or other structural issues that could compromise the stability of the entire structure. Timely detection and correction of such deviations are critical to preventing serious accidents.

Various methods are employed to determine the verticality of structures, including observations using global navigation satellite systems (GNSS), three-dimensional ground-based geodetic measurements with total stations, theodolites, inclinometers, and laser scanning systems (Gao et al., 2016; Li et al., 2014; Roberts et al., 2004; Song et al., 2010; Wang et al., 2018; Wang, 2013; Widerski & Kurałowicz, 2014). However, when additional information about an object's geometric parameters is required, or when high-precision measurements are needed in complex or remote areas, alternative approaches become necessary.

Laser scanning methods offer high detail and accuracy in generating three-dimensional models, they require significant financial and technical resources, limiting their application in engineering tasks. Moreover, these methods do not always guarantee the positional accuracy necessary for certain applications. In contrast, GNSS technologies are effective over large areas but typically provide accuracy within a few meters, which is inadequate for precise engineering solutions.

With advancements in technology, unmanned aerial vehicles (UAVs) have quickly integrated into remote monitoring of infrastructure facilities, becoming essential tools in this field, with their use continuing to grow (Ham et al., 2016; Duque et al., 2018; Albeaino et al., 2019; Onososen et al., 2023; Lobanov et al., 2023; Lobanov et al., 2024). UAVs have proven

DOI: 10.1201/9781003518518-30

effective in determining the tilt of towers by analyzing images captured during aerial surveys (Achille et al., 2015; Liang & Liu, 2016; Lu et al., 2022; Markiewicz et al., 2017).

However, integrating these modern methods with other technologies is equally important, as it can significantly enhance the capabilities of remote diagnostics. In particular, improved data processing algorithms and automation of processes increase the accuracy and efficiency of monitoring infrastructure conditions.

2 METHODOLOGY

To determine the verticality of the lower tier of the Kyiv TV tower, a series of experimental measurements were conducted using a DJI Mavic 3 Enterprise unmanned aerial vehicle (UAV) and an M2Pro RTK GNSS receiver. The purpose of the study was to ensure high-precision capture of the object's spatial coordinates for further processing and analysis.

The survey was carried out considering optimal parameters, including selecting the appropriate flight altitude, adjusting the photography angle, and accounting for weather conditions to minimize errors.

The data was collected through a series of overflights of the TV tower at the shortest possible distance from the object, which increased the resolution and clarity of the images.

One of the key tools for achieving high accuracy in geospatial coordinates for each survey position is the use of RTK (Real-Time Kinematic). The RTK receiver provides real-time correction of satellite signals, enabling highly accurate coordinates of the survey points with centimeter-level precision.

Once the data from the RTK receiver was collected, it was processed using the photogrammetry method. This technique enables the creation of a three-dimensional point cloud based on images captured from multiple viewpoints. The 3DF Zephyr software was used to process the data and generate a 3D model of the TV tower (Toldo, R, et al., 2015). The key feature was the use of 3D GNSS coordinates of each input image as Constraints and Check Points. The model reconstruction process is fully automated and does not require additional adjustments or the use of coded targets for scaling (Figure 1). This approach allows for detailed and accurate reproduction of the spatial characteristics of the object, which is crucial for precise monitoring and analysis of the condition of infrastructure facilities. Based on the resulting 3D model, the central trunk of the TV tower was identified, serving as the key reference point for verticality analysis.

To determine the position of the central axis of the TV tower, the point cloud was segmented through photogrammetric analysis. This process involved dividing the point cloud into individual segments, each representing a specific section of the structure. For each segment, the least-squares method was employed to approximate the geometric shape of the section, minimizing the sum of squares of deviations between the points of a circle. This approach enabled the accurate determination of centers for every section (Figure 2, a). This provided precise coordinates of the cross-sectional centers, reflecting their actual position in 3D space. These center coordinates were used in subsequent calculations of the tower's verticality.

3 RESULTS

To assess the verticality of the lower tier of the TV Tower, the centers of the elevator shaft sections were calculated using the least-squares method from dense point data obtained through photogrammetry software, with a height interval of one meter.

From the analysis of the central axis deviation graphs, a linear deviation is observed between +15 and +40 meters, indicating the absence of bends within this height range. However, between +40 and +80 meters, a bend of the elevator shaft is detected between +50 and +65 meters (Figure 2, b), where the X coordinate corresponds to the east-west direction and the Y coordinate represents the north-south direction. The bend can be caused by damage at +58 meters of the lattice structure of the TV tower.

Figure 1. 3D model of the low tier of the Kyiv TV tower.

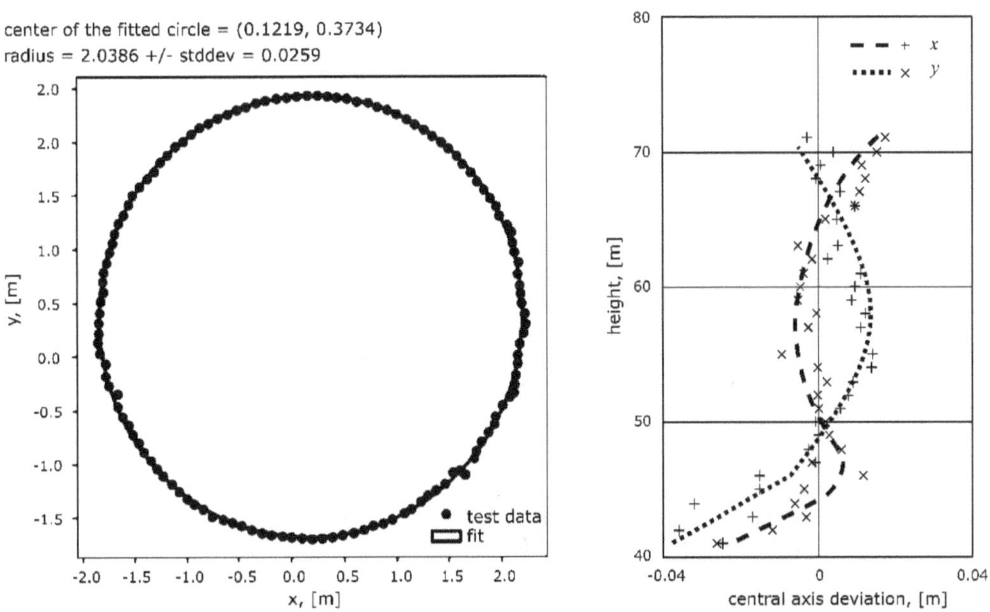

a) b)

Figure 2. Assessment of the vertical alignment of the lower tier of the TV Tower: a) calculation of the center coordinates of an elevator shaft section, b) relative coordinates of the central axis with respect to height, section +40 to +80.

The distribution of points forming a circle in each section (Figure 2, a), together with the tendency of central axis deviation throughout the shaft height (Figure 2, b), indicates photogrammetry accuracy of at least the centimeter level.

4 CONCLUSIONS

This study demonstrates that the integration of unmanned aerial vehicles (UAVs) with photogrammetry and RTK modules enables the creation of highly accurate point clouds and detailed three-dimensional models of large structures, such as TV towers. The application of RTK data and bundle adjustment algorithms enhances the precision of photogrammetry measurements, achieving centimeter-level accuracy for structures over one hundred meters in height.

The analysis of the TV tower revealed deviations in verticality. These findings validate the effectiveness of the combined UAV-RTK-photogrammetry approach for infrastructure monitoring and verticality assessment. This methodology is crucial for maintaining the safety and structural integrity of tall infrastructure, highlighting its potential as a reliable tool for ongoing structural health monitoring.

ACKNOWLEDGEMENT

This research was conducted within project 022.01/0095 "Development of technology for remote diagnosis of damaged large-scale objects based on the use of unmanned aerial vehicles (UAVs) and photogrammetry" funded by the National Research Foundation of Ukraine in the framework "Science for the Recovery of Ukraine in the War and Post-War Periods".

REFERENCES

Achille, Cristiana & Adami, Andrea & Chiarini, Silvia & Cremonesi, Stefano & Fassi, Francesco & Fregonese, Luigi & Taffurelli, Laura. (2015). UAV-Based Photogrammetry and Integrated Technologies for Architectural Applications—Methodological Strategies for the After-Quake Survey of Vertical Structures in Mantua (Italy). Sensors. 15. 15520–15539. 10.3390/s150715520.

Albeaino, G., Gheisari, M., & Franz, B. W. (2019). A systematic review of unmanned aerial vehicle application areas and technologies in the AEC domain. Journal of Information Technology in Construction (ITcon), 24, 381–405. https://www.itcon.org/2019/20.

Duque, L., Seo, J., and Wacker, J. (2018). Synthesis of Unmanned Aerial Vehicle Applications for Infrastructures. Journal of Performance of Constructed Facilities, 32(4), 04018046. https://doi.org/10.1061/(ASCE)CF.1943-5509.0001185.

Gao, Yang & Lin, Jiarui & Linghui, Yang & Zhu, Jigui. (2016). Development and calibration of an accurate 6-degree-of-freedom measurement system with total station. Measurement Science and Technology. 27. 10.1088/0957-0233/27/12/125103.

Ham, Y., Han, K. K., Lin, J. J., and Golparvar-Fard, M. (2016). Visual monitoring of civil infrastructure systems via camera-equipped Unmanned Aerial Vehicles (UAVs): a review of related works. Visualization in Engineering, 4(1), 1. https://doi.org/10.1186/s40327-015-0029-z.

Liang, Chen & Liu, Zhengjun & Jianguo, Qian. (2016). Airborne LIDAR point cloud tower inclination judgment. IOP Conference Series: Earth and Environmental Science. 46. 012013. 10.1088/1755-1315/46/1/012013.

Lobanov, L.M. & Stelmakh, D.I. & Savitsky, V.V. & Diadin, V.P. & Shutkevych, O.P. & Kozachek, A. G. (2023). Remote assessment of damage to Kyiv TV tower based on the application of aerial photography and photogrammetry method. Tehničeskaâ diagnostika i nerazrušaûŝij kontrol'. 2023. 16–20. 10.37434/tdnk2023.03.03.

Lobanov, Leonid & Stelmakh, Dmytro & Savitsky, Viktor & Diadin, Vitalii & Shutkevych, Oleksii & Shkurat, Inna. (2024). Damage detection and analysis using unmanned aerial vehicles (UAVs) and photogrammetry method. Procedia Structural Integrity. 59. 43–49. 10.1016/j.prostr.2024.04.008.

Lu, Zhumao & Gong, Hao & Jin, Qiuheng & Hu, Qingwu & Wang, Shaohua. (2022). A Transmission Tower Tilt State Assessment Approach Based on Dense Point Cloud from UAV-Based LiDAR. Remote Sensing. 14. 408. 10.3390/rs14020408.

Markiewicz, Jakub & Zawieska, Dorota & Podlasiak, Piotr. (2017). Comparing multi-source photogrammetric data during the examination of verticality in a monumental tower. ISPRS - International Archives of the Photogrammetry, Remote Sensing and Spatial Information Sciences. XLII-2/W3. 475–480. 10.5194/isprs-archives-XLII-2-W3-475-2017.

Onososen, Adetayo & Musonda, Innocent & Onatayo, Damilola & Tjebane, Motheo & Saka, Abdullahi & Fagbenro, Rasaki. (2023). Impediments to Construction Site Digitalization Using Unmanned Aerial Vehicles (UAVs). Drones. 7. 10.3390/drones7010045.
Roberts, Gethin & Meng, Xiaolin & Dodson, Alan. (2004). Integrating a Global Positioning System and Accelerometers to Monitor the Deflection of Bridges. Journal of Surveying Engineering-ASCE. 130. 10.1061/(ASCE)0733-9453(2004)130:2(65).
Toldo, R.; Gherardi, R., Farenzena, M., Fusiello, A. (2015) Hierarchical structure-and-motion recovery from uncalibrated images, Computer Vision and Image Understanding, Volume 140, p.127–143, 10.1016/j.cviu.2015.05.011.
Song, Jiming & Song, Huasong & Wang, Yiwen. (2010). Methods of measurement at the Special Terrain tower tilted rate. Power System Technology. 34(12): 119–222.
Wang, Ruigang. (2013). On optimized tower tilt measurement and preventive measures. Management & Technology. 2013 (12).
Wang, S.; Du, Y.; Sun, J.; Fang, Q.; Weng, Y.; Ma, L.; Zhang, X.; Wu, J.; Qin, Q.; Shi, Q. (2018). Transmission tower deformation and tilt detection research status. Telecom Power Technol. 35, 91–92.
Widerski, T. & Kurałowicz, Z. (2014). Geodesic monitoring of tower and mast structures. Reports on Geodesy. 411–417.

Welding and Related Technologies – Krivtsun et al. (Eds)

Applying neural network technologies for quality control in robotic surfacing

A.S. Novodranov & E.V. Shapovalov
E.O. Paton Electric Welding Institute of the National Academy of Sciences of Ukraine, Kyiv, Ukraine

ABSTRACT: 3D printing technology is currently actively used to manufacture parts in various industries. The production of metal parts using additive deposition may be accompanied by the appearance of surface defects in the deposited layers, namely pores and cracks. Using a robotic complex based on an anthropomorphic robot with an integrated machine vision system increases the surfacing process's quality and productivity. The hardware component of the machine vision system consists of a high-speed GigE camera and a controller. The software uses the YOLO-NAS neural network architecture to detect and classify defects automatically after the completion of the deposition process of the corresponding layer. The quantitative characteristics of the quality of recognition and classification using a trained neural network model are given. It is recommended that anthropomorphic robots and machine vision systems based on neural network technologies be used for manufacturing parts for a responsible purpose using WAAM technology.

1 INTRODUCTION

Currently, the production of parts using 3D printing technology is rapidly spreading in various branches of industrial production. The WAAM (Wire Arc Additive Manufacturing) technology for manufacturing complex-shaped metal parts using layer-by-layer surfacing with subsequent minimal mechanical processing is also becoming widespread (Shah et al., 2023). WAAM additive surfacing systems use solid and powder welding wire as consumables for this process. Using a robotic complex solves the problem of fully automating the production process of manufacturing a part. A significant advantage of automation is the minimisation of the influence of the human factor on the production process, even to its complete exclusion. The leading causes of surfacing defects include incorrectly selected surfacing technology, malfunctioning welding equipment, low-quality or incorrectly selected welding materials, and other factors. The probability of the formation of surface defects primarily depends on the surfacing technology's perfection and the quality of the welding materials. In the case of untimely detection of such defects, the formation of similar defects may occur in the following layers, which may cause a deterioration of the product's operational characteristics. The modern development of computer technologies and artificial intelligence, namely machine learning, allows the timely detection and classification of surface defects typical for a specific production. The goal of this paper is to investigate the effectiveness of using machine vision systems based on neural network technologies in the additive robotic surfacing of metal parts.

2 PROBLEM STATEMENT AND METHODOLOGY

The presence of welding spatter during WAAM surfacing leads to negative consequences. During automated surfacing with the help of a robotic complex, the torch nozzle gradually becomes contaminated from the inside due to splashes. This leads to the deterioration of the gas protection of the welding zone, which, in turn, can cause the appearance of pores in the metal of the deposited

DOI: 10.1201/9781003518518-31

layer. In part, this problem can be solved by using an automatic torch cleaning station, which will significantly reduce the probability of the appearance of pores but does not guarantee their complete absence. As is known, the presence of cracks after surfacing, for example, wear-resistant layers with high-strength wire, can most likely destroy the product during an operation. The deposition of wear-resistant layers is accompanied by an increase in the hardness and strength of the deposited layer, but at the same time, it makes it more brittle. This is due to a significant amount of carbon and alloying elements, namely chromium, molybdenum, and vanadium, in the welding wire. In typical robotic surfacing systems, the operator performs the function of visual quality control of the deposited layer, which can lead to the omission of some defects. In manufacturing parts using 3D printing, the number of NDT methods that can be used is limited. The authors proposed using a visual method of non-destructive control after the additive deposition of a particular layer. Equipping the robotic complex with a machine vision system and using layer-by-layer visual NDT can significantly improve the quality of manufactured parts. Among the advantages of the visual non-destructive testing method are high speed, no need for contact with the object of control, and the possibility of automation. It can be noted, however, that the detection of only visible surface defects is a disadvantage of this method (Dwivedi et al., 2018).

3 ROBOTIC NEURAL NETWORK SYSTEM FOR VISUAL QUALITY CONTROL

The automation of the layer deposition process was carried out using a robotic complex based on the Fanuc ARC Mate 100iC/7L welding robot (see Figure 1a). The use of the robot allows you to reproduce the pre-programmed trajectory of the torch accurately, as well as control the time delays between the depositions of layers. At the same time, the positioning accuracy of the torch reaches ± 0.03 mm. With the help of the machine vision system (see Figure 1b), the defectiveness of the deposited layer was checked. Typical surface defects that occur in the process of additive surfacing include pores and cracks.

The hardware part of the machine vision system consists of a Basler acA1920-48gc colour high-speed Gigabit Ethernet camera, rigidly fixed to the torch on a unique mount, and an image processing controller. The camera resolution is 1920 x 1200 (2.3 MP), and the frame rate is 50 fps. The camera has a lens (HM1214MP5) with a constant focal length of 12 mm and a set of light filters. The image processing controller performs the following functions: brightness and contrast normalisation, noise removal, and resizing; identification and classification of surface defects of deposited layers; determination of the physical dimensions of the detected defect; exchange of data with the welding robot for the formation of welding programmes according to a particular protocol.

The software implementation of the proposed system for detecting and classifying surface defects in deposited layers consists of the following main functional blocks: video stream capture module; module for extracting frames from a video stream; module for pre-processing frames; feature selection module; module for classification of detected surface defects; module for determining the size and position of the defect. The video stream capture module is designed to record images of the deposited layer surface in real time. The frame extraction module is designed to extract individual frames from a video stream. The pre-processing module processes received RAW images, namely resizing, normalising brightness and contrast, removing noise, and others. The module for selecting signs of surface defects based on a neural network performs the selection of characteristic signs of defects on images in automatic mode. The classification module determines the class of surface defects, namely pores or cracks, based on the selected features. The software library of computer vision algorithms with the open source code OpenCV was used to determine the geometric dimensions and position of the defective area. Determining the geometric dimensions and location parameters of the defective areas relative to the boundaries of the deposited layer is performed with the help of a known standard to be installed next to the deposited product. This simplification of determining the object's physical dimensions in the image is due to the rigid fixation of the camera on the robot torch. Therefore, by calibrating the camera relative to the robot's TCP (Tool Centre Point), a simple transition between the robot's coordinate systems and the coordinates obtained after image processing is achieved.

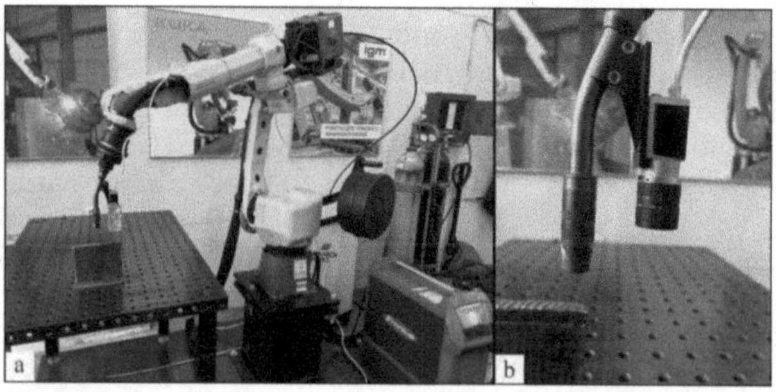

Figure 1. Robotic system: a – general view of the robotic system; b – visual inspection module.

The supervised machine learning method was used to ensure the training of the YOLO-NAS neural network model (Terven et al., 2023). Given the limited number of images for training, the data augmentation method was applied, which consists of creating modified versions of images using geometric and colour transformations. As a result, 547 images with a resolution of 640 × 640 pixels were obtained. The dataset contains defects of two classes, namely 485 weld pore images and 62 crack images. The next step was to label the defective areas in the images and indicate their class using the open-access software LabelImg. 80% of the total data set was the training sample, and 10% each was the test and verification sample. This is necessary to obtain an objective assessment of the model's performance. The duration of the learning process was 300 epochs. With the help of the PyTorch library in the Python programming language, the neural network model was trained. Graphs of average accuracy and losses are obtained, which are designed to analyse the learning process, as well as to analyse the balance of various aspects of model training. Also, these graphs help to determine the moment of the start of the retraining process, that is when the model works correctly on the training data, but the result is meager on the new data. A decrease in the graph of the average accuracy after a long stabilisation may indicate the beginning of retraining of the model.

4 RESULTS

Estimating the performance of a neural network model begins with obtaining true and predicted labels for the test data set generated after the training process. The next step is to create a confusion matrix of size N × N, where N is the number of classes. For its construction, it is necessary to determine the number of unique classes and obtain a set of predicted and true labels. The main diagonal of this matrix displays the number of correct predictions for each class. As a result of processing the matrix of inconsistencies, the following components are extracted, which show the result of comparing the true and predicted labels: True Positives (TP); False Positives (FP); True Negatives (TN); False Negatives (FN). The next step is the calculation of metrics (Hossin & Sulaiman, 2015) for evaluating the effectiveness of the detection and classification model. Firstly, the Precision metric is calculated, which shows the proportion of correctly predicted positive cases among all positive cases and is mathematically expressed as follows:

$$\text{Precision} = \frac{\text{TP}}{\text{FP} + \text{TP}} \tag{1}$$

Next, the Recall metric is calculated, which shows the share of precisely predicted positive cases among all real positive cases and is presented in the following form:

$$\text{Recall} = \frac{\text{TP}}{\text{TN} + \text{TP}} \tag{2}$$

Based on the values of the Precision and Recall metrics, the F_{score} metric is calculated. The mathematical calculation of the metric can be represented as follows:

$$F_{score} = 2 \times \frac{\text{Precision} \times \text{Recall}}{\text{Precision} + \text{Recall}} \tag{3}$$

The results of the model for recognising and classifying surface defects of the deposited layer are presented in Figure 2. The detection result, which contains the coordinates and dimensions of the defective zone, the confidence value and the determined class of the defect, is presented in JSON format. Information on the detection and classification of surface defects presented in this way is used to calculate the physical dimensions of the object and determine its position relative to the deposited layer. Quantitative indicators of the training quality of the YOLO-NAS model, as well as the detection and classification of surface defects in deposited layers in comparison with YOLOv8, are shown in Table 1.

Table 1. Quantitative indicators of the quality of defect detection and classification.

Model	Defect type	mAP, %	Precision, %	Recall, %	F_{score}
YOLO-NAS-M	Crack		87.50	87.50	0.91
	Pores	90.15	94.31	88.92	
	All		94.11	88.90	
YOLOv8	Crack		77.78	99.40	0.90
	Pores	89.56	93.56	87.90	
	All		93.04	88.19	

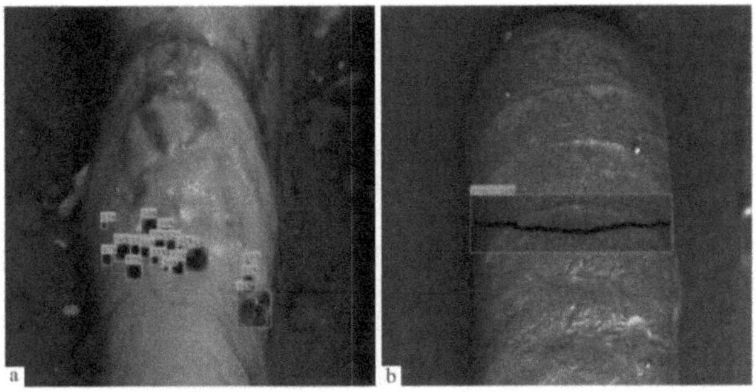

Figure 2. Recognised defects of the deposited layer: a – accumulation of pores; b – transverse crack.

5 CONCLUSIONS

1. A visual non-destructive testing method is proposed for searching for surface defects during WAAM, which makes it possible to detect pores and cracks in each deposited layer.
2. For automated detection of surface defects in deposited layers, it is proposed to use a convolutional neural network with the YOLO-NAS architecture, which has satisfactory quality indicators for defect detection and classification.
3. Convolutional neural networks have been shown to be highly effective for searching and classifying surface defects such as pores and cracks.

REFERENCES

Dwivedi, S.K., Vishwakarma, M, Soni, A. 2018. Advances and Researches on Non Destructive Testing: A Review. *Materials Today: Proceedings*, 5(2), 3690–3698.

Hossin M, Sulaiman M.N. 2015 A Review on Evaluation Metrics for Data Classification Evaluations. *International Journal of Data Mining & Knowledge Management Process*, 5, 1–11.

Shah A.H., Aliyev R., Zeidler H., Krinke S. 2023 A Review of the Recent Developments and Challenges in Wire Arc Additive Manufacturing (WAAM) Process. *J. Manuf. Mater. Process*, 7, 97.

Terven J., Córdova-Esparza D.M., Romero-González J.A. 2023 A Comprehensive Review of YOLO Architectures in Computer Vision: From YOLOv1 to YOLOv8 and YOLO-NAS. *Machine Learning and Knowledge Extraction*, 5(4), 1680–1716.

Welding and Related Technologies – Krivtsun et al. (Eds)

Method for analysis of periodically non-stationary properties of vibrations for a gas-turbine engine with unbalanced rotor

I.M. Javorskyj
Karpenko Physico-mechanical Institute of NAS of Ukraine, Ukraine

R.M. Yuzefovych
Karpenko Physico-mechanical Institute of NAS of Ukraine, Ukraine
Lviv Polytechnic National University Lviv, Ukraine

O.V. Lychak
Karpenko Physico-mechanical Institute of NAS of Ukraine, Ukraine

Yu.I. Torba
State Enterprise Zaporizhzhia Machine-Building Design Bureau "PROGRESS" named after Academician O. H. Ivchenko , Zaporizhzhia, Ukraine

Ye.V. Sbrodov
State Enterprise Zaporizhzhia Machine-Building Design Bureau "PROGRESS" named after Academician O. H. Ivchenko , Zaporizhzhia, Ukraine
National University "Zaporizhzhia Polytechnic" Zaporizhzhia, Ukraine

ABSTRACT: Method for the vibrations analysis of the gas-turbine engine on the basis of their model in the form of periodically non-stationary random processes (PNRPs) was presented. It is assumed that the power spectral density of the raw signal is mixed. The least square method is used for discovering of the hidden periodicities in the vibration signal. The frequencies of the deterministic basic harmonics are estimated as well as their amplitudes. It is revealed that for the case of unbalanced rotor the mean is the time periodical function, basic frequency of which is close to rotation frequency. Increasing of the amplitudes of hidden components reveals the appearance of the distributed fault.

1 INTRODUCTION

It is known in literature (Djaidir et al. 2017, Fábry & Českovič 2017, Pérez-Ruiz et al. 2024, Rath et al. 2022, Grządziela 2006, Zaidan et al. 2015, Stamatis 2011), that the vibrations of the gas-turbine engine with unbalanced rotor are characterized by the peak values of their amplitude or power spectra at the point, which is defined by the first harmonic of rotor rotation frequency. The amplitude spectrum as a rule is found using the Fourier transform (FT) of the raw signal. The power spectrum is the result of applying of the periodogram or Blackman-Tukey spectral estimator. It should be noted, that a lot of known techniques conforms to the different representations of vibrations (Heng et al. 2009, Soleimani et al. 2021, Lakshminarasimha et al. 1994, Marinai et al. 2004, Napolitano 2020). Using FT, investigators have found that time series are the segments of the deterministic oscillations, while calculating of the power spectra they considered them as realizations of the stationary random processes. If we suppose that the experimental data are described by the sum of harmonics and random process, then both procedures are inadequate to this representation and they result in inconsistent sequels. In these cases the revealing of the harmonics and estimating of their amplitudes must be carried out proceeding from the vibration signal model in the form of the PNRP. This approach is used in the present work.

DOI: 10.1201/9781003518518-32

2 AMPLITUDE AND POWER SPECTRUM OF THE RAW SIGNAL

Suppose that the analyzed signal $\xi(t)$ is the sum of stationary random process $\eta(t)$ and almost periodical function

$$s(t) = \sum_{k=-M}^{M} c_k e^{i\omega_k t},$$

where M is number of harmonics, ω_k and c_k are their frequencies and amplitudes. The signal sample values $\xi(nh)$ are acquired for $n \in [-K, K]$, $2K = \Theta/h$, Θ is the realization length, h is sampling interval. Then for the signal discrete time FT (Koopmans 1974, Kay 1988) we have:

$$\hat{c}_\xi(\omega) = \frac{h}{2\pi} \sum_{n=-K}^{K} [\eta(nh) + s(nh)] e^{-i\omega_k nh}. \tag{1}$$

The mathematical expectation of (1) is equal to:

$$E\hat{c}_\xi(\omega) = \frac{h}{2\pi}\left[m_\eta \sum_{n=-K}^{K} e^{-i\omega_k nh} + \sum_{k=-M}^{M} c_k \sum_{n=-K}^{K} e^{-i(\omega-\omega_k)nh}\right],$$

where $m_\eta = E\eta(nh)$ is a mean of $\eta(nh)$. Assuming that $h \leq \pi/\omega_M$ and using the formula for the sum of geometric progression we obtain

$$\sum_{n=-K}^{K} e^{-i\omega nh} = 1 + \sum_{n=1}^{K}\left(e^{i\omega nh} + e^{-i\omega nh}\right) = \varphi(\omega, K),$$

where $\varphi(\omega, K) = \sin\omega\left(K + \frac{1}{2}\right)h / \sin\omega\frac{n}{2}$. Hence $E\hat{c}_\xi(\omega) \approx \frac{h}{2\pi}\left[m_\eta\varphi(\omega, K) + \sum_{k=-M}^{M} c_k\varphi(\omega - \omega_k, K)\right]$.
Since $\varphi(0, K) = 2K + 1$, then for the large K we get:

$$E\hat{c}_\xi(\omega) \approx (\Theta/\pi)c_k. \tag{2}$$

It means that we cannot determine the amplitudes of harmonics for signal $\xi(t) = \eta(t) + s(t)$ using FT, because the values (2) are different according to realization length θ and $E\hat{c}(\omega_k) \to \infty$, as $\theta \to \infty$. Herewith the spectral properties of the random process in general do not characterize. These properties can be described using the higher moment functions of the random process.

The first well-known results in this field were obtained by the N. Wiener and A. Khinchin (Koopmans 1974). They introduced concept of the power spectral density as the FT of the covariance function $R_\xi(u) = E\,\overset{\circ}{\xi}(t)\,\overset{\circ}{\xi}(t+u)$, $\overset{\circ}{\xi}(t) = \xi(t) - m_\xi(t)$:

$$f_\xi(\omega) = \frac{1}{2\pi} \int_{-\infty}^{\infty} R_\xi(u) e^{-i\omega u} du. \tag{3}$$

To estimate the power spectral density on the basis of experimental data the correlogram method of Blackman-Tukey can be used:

$$\hat{f}_\xi(\omega) = \frac{1}{2\pi} \int_{-u_m}^{u_m} \hat{R}_\xi(u) k(u) e^{-i\omega u} du,$$

here $\hat{R}_\xi(u)$ is the estimator of covariance function:

$$\hat{R}_\xi(u) = \frac{1}{K}\sum_{n=1}^{K} [\xi(nh) - \hat{m}][\xi(nh+u) - \hat{m}], \; \hat{m} = \frac{1}{K}\sum_{n=1}^{K} \xi(nh), \tag{4}$$

$k(u)$ is the covariance window, u_m is cut-of point of the correlogram.

The discrete estimator of (3) has the form:

$$\hat{f}_{\xi}(\omega) = \frac{h}{2\pi}\sum_{n=-L}^{L}\hat{R}_{\xi}(nh)k(nh)e^{-i\omega nh}, \tag{5}$$

where $L = u_m/h$. For the asymptotic value of the mathematical expectation of (5) we find:

$$E\hat{f}_{\xi}(\omega) = \frac{h}{2\pi}\sum_{n=-L}^{L}R_{\xi}(nh)k(nh)e^{-i\omega nh}.$$

The time-averaged covariance function of the signal (its covariance function in the stationary approximation) is equal to:

$$\hat{R}_{\xi}(nh) = \hat{R}_{\eta}(nh) + \sum_{k=-M}^{M}|c_k|^2 e^{i\omega_k nh}. \tag{6}$$

Assume for simplicity that the covariance window is rectangular: $k(u) = \begin{cases} 1, & |u| \leq u_m, \\ 0, & |u| > u_m. \end{cases}$
Then (Javorskyj et al. 2021)

$$E\hat{f}_{\xi}(\omega) = \int_{-\infty}^{\infty} f_{\eta}(\omega_1)\lambda(\omega - \omega_1)d\omega_1 + \frac{h}{2\pi}\sum_{k=-M}^{M}|c_k|^2\varphi(\omega - \omega_k, L).$$

Since $\varphi(0, L) = 2L + 1$, then the peak values $E\hat{f}_{\xi}(\omega)$ are proportional to u_m. Therefore searching for these peaks we can only find the frequencies of the harmonic components. To determine their powers (and amplitudes) we must use another approach.

3 HARMONIC ANALYSIS OF THE DETERMINISTIC OSCILLATIONS

The estimation of the basic frequencies of the deterministic components is the first issue of the vibration harmonic analysis. To find the basic frequencies, the special functionals can be used (Javorskyj et al. 2021, Javorskyj et al. 2017), that are similar to the PNRP coherent or component statistics for mean function estimators with the except that test basic frequency is inserted to replace the true period. These functionals have extreme values at the points, which are asymptotically unbiased and consistent basic frequency estimators. To improve the efficiency of estimating of basic frequencies, the least square (LS) method was proposed in (Javorskyj et al. 2022). As the signal realization length increases, the LS functional for determining of the basic frequencies of the mean function quickly tends to

$$\hat{F}_1(f) = \frac{1}{2}\sum_{k=1}^{L_i}\left[\left[\hat{m}_k^c(f)\right]^2 + \left[\hat{m}_k^s(f)\right]^2\right], \tag{7}$$

where

$$\begin{Bmatrix} \hat{m}_k^c(f) \\ \hat{m}_k^s(f) \end{Bmatrix} = \frac{2}{2K+1}\sum_{n=-K}^{K}\xi(nh)\begin{Bmatrix} \cos(2k\pi fnh) \\ \sin(2k\pi fnh) \end{Bmatrix},$$

and 2Θ is the realization length, $h = \frac{\Theta}{K}$ is the sampling step and L_i is number of harmonics for each component, f is a test frequency. The maximum points $\hat{f}_0^{(i)}$ of (7) is asymptotically unbiased and consistent estimator for the mean basic frequency. The functional (7) at the maximum point $f = \hat{f}_0^{(i)}$ is close to the sum of time-averaged power of the chosen harmonics:

$$\hat{F}_1\left(\hat{f}_0^{(i)}\right) = \hat{P}_t^{(d_i)} = \frac{1}{2}\sum_{k=1}^{L_i}\left[\left[\hat{m}_k^c\left(\hat{f}_0^{(i)}\right)\right]^2 + \left[\hat{m}_k^s\left(\hat{f}_0^{(i)}\right)\right]^2\right].$$

The quantities $\hat{m}_k^c\left(\hat{f}_0^{(i)}\right)$ and $\hat{m}_k^s\left(\hat{f}_0^{(i)}\right)$ are asymptotically unbiased and consistent estimators of the Fourier coefficients for the mean function of each component. Hence, we can form the mean function estimator for each component:

$$\hat{m}_i(t) = \sum_{k=1}^{L_i}\left[\hat{m}_k^c\left(\hat{f}_0^{(i)}\right)\cos 2k\pi\hat{f}_0^{(i)}t + \hat{m}_k^s\left(\hat{f}_0^{(i)}\right)\sin 2k\pi\hat{f}_0^{(i)}t\right]. \tag{8}$$

The functions (8) describe the deterministic components of vibrations.

4 CONCLUSION

Knowing $\hat{f}_0^{(i)}$ we calculate the amplitudes of the respective harmonics $A\left(k\hat{f}_0^{(i)}\right) = \left[\left[m_k^c\left(\hat{f}_0^{(i)}\right)\right]^2 + \left[m_k^s\left(\hat{f}_0^{(i)}\right)\right]^2\right]^{\frac{1}{2}}$. The ratio of the amplitudes for the different engine state is equal to: $I_1 = A^{(f)}\left(k\hat{f}_0^{(2)}\right)/A^{(h)}\left(k\hat{f}_0^{(2)}\right)$. We can choose this ratio as well indicator for the description of the rotor state.

REFERENCES

Djaidir, B.; Hafaifa, A. & Kouzou A. 2017. Faults detection in gas turbine rotor using vibration analysis under varying conditions. *J. Theoretical and Applied Mechanics* 55(2): 393–406.

Fábry S. & Češkovič M. 2017. Aircraft gas turbine engine vibration diagnostics. *Review Articles* 5(4): 24–28.

Grządziela, A. 2006. Analysis of vibration parameters of ship gas turbine engines. *Marine Engineering* 2: 22–26.

Heng, A.; Zhang, S.; Tan, A.C. & Mathew, J. 2009. Rotating machinery prognostics: state of the art, challenges and opportunities. *Mech. Syst. Signal. Process.* 23: 724–739.

Javorskyj, I.; Matsko, I.; Yuzefovych, R.; Lychak, O. & Lys, R. 2021. Methods of hidden periodicity discovering for gearbox fault detection. *Sensors* 21: 6138.

Javorskyj, I.; Yuzefovych, R.; Matsko, I. & Zakrzewski, Z. 2022. The least square estimation of the basic frequency for periodically non-stationary random signals. *Dig. Sign. Proc.* 122: 103333.

Javorskyj, I.; Yuzefovych, R.; Matsko, I.; Zakrzewski, Z. & Majewski, J. 2017. Coherent covariance analysis of periodically correlated random processes for unknown non-stationarity period. *Dig. Signal Process.* 65: 27–51

Kay, S.M. 1988. *Modern Spectral Estimation: Theory and Application.* New Jersey: Prentice-Hall, Englewood.

Koopmans, L.H. 1974. *The spectral analysis of time series [by] L.H. Koopmans.* New York: Academic Press.

Lakshminarasimha, A.N.; Boyce, M.P. & Meher-Homji, C.B. 1994. Modeling and analysis of gas turbine performance deterioration. *J. Eng. Gas Turbines Power* 116: 46–52.

Marinai, L., Probert, D. & Singh, R. 2004. Prospects for aero gas-turbine diagnostics: a review. *Appl. Energy* 79(1): 109–126.

Napolitano, A. 2020. *Cyclostationary processes and time series: Theory, applications, and generalizations.* Elsevier, Academic Press.

Pérez-Ruiz, J.L.; Luis, J.; Tang, Yu; Loboda, I. & Miró-Zárate, L. 2024. An Integrated Monitoring, Diagnostics, and Prognostics System for Aero-Engines under Long-Term Performance Deterioration. *Aerospace* 11(3): 217, 11020217.

Rath, N.; Mishra, R.K. & Kushari, A. 2022. Aero engine health monitoring, diagnostics and prognostics for condition-based maintenance: an overview. *Int. J. Turbo Jet Engines.*

Soleimani, M.; Campean, F. & Neagu, D. 2021. Diagnostics and prognostics for complex systems: a review of methods and challenges. *Qual. Reliab. Engng. Int.* 37: 3746–3778.

Stamatis, A.G. 2011. Evaluation of gas path analysis methods for gas turbine diagnosis. *J. Mech. Sci. Technol.* 25: 469–477.

Zaidan, M.; Relan, R.; Mills, A. & Harrison, R. 2015. Prognostics of gas turbine engine: An integrated approach. *Expert Systems with Applications* 42 (22): 8472–8433.

Session 7. New structural and functional materials, nanomaterials, composites

Welding and Related Technologies – Krivtsun et al. (Eds)

Development of chemical composition and heat treatment of pearlite steel for high-strength railway rails

O.I. Babachenko, G.A. Kononenko, R.V. Podolskyi & O.A. Safronova
Iron and Steel Institute of Z.I. Nekrasov of National Academy of Sciences of Ukraine, Dnipro, Ukraine

ABSTRACT: Currently, changes to the requirements for the manufacture of railway rails cause the need for new chemical compositions and heat treatment. Boron and calcium micro-alloying is a rather promising direction in the development of railway rails in accordance with the DSTU EN 13674:2018 standard. In the work, steel with carbon at the level of 0.9% was produced with boron and calcium microalloying. A study of the influence of the cooling rate on the structural properties was carried out, and the recommended cooling parameters and the dependence of the cooling rate on the interlaminar distance of pearlite and bulk hardness were established. The following chemical composition is recommended for experimental and industrial testing of railway rail manufacturing technology: C = 0.90%, Si = 0.39%, Mn = 0.89%, Ca = 0.0006%; B=0.0003% and recommended heat treatment parameters: accelerated cooling of the rail head at a rate of 5.1±0.2°C/s with tempering at 200°C.

1 INTRODUCTION

Rails are one of the most important elements of a railway track. To ensure high efficiency and reliability of railway transport, it is necessary to increase the stability of railway rails. Currently, progressive requirements are in force in Ukraine, regulated by DSTU EN 13674:2018 for high-strength pearlite rails of the R400HT category. Among them, the most difficult to achieve are the requirements for hardness (370 HB at a depth of 20 mm from the surface of the rail head). The current technology made it possible to stably fulfill the requirements for rails of the highest category according to DSTU 4344:2004 - hardness of at least 321 HB at a depth of 11 mm. For many years, this standard was the only valid standard in Ukraine before the introduction of DSTU, which is identical to the EN 13674. One of the ways to solve the problem is to improve the existing chemical composition of steel, which will ensure increased hardenability in the process of heat treatment in order to form a dispersed pearlite without the formation of martensite and cementite.

Today, according to the authors of the works (Zhou et al. 2023, Skoblo T.S. 2006), it is possible to achieve a significant increase in the resistance of the rail against abrasion by producing rails from super-eutectoid steels, bainite class steels and microalloying high-carbon steels. Currently, pre-eutectoid steels are often used. The effectiveness and reliability of this approach is confirmed by the introduction of the chemical composition of steel with a carbon content range of 0.88…1.07% into the regulatory and technical documentation. This made it possible to expand the potential for the use of post-eutectoid steels without going beyond the requirements of the regulations. Application of microalloying is possible in several directions. One of them is the formation of dispersed carbonitrides in high-carbon rail steel (Poschitz & Kolesov 1998, Mujica Roncery et al. 2011, Boettinger et al. 2002), which provides a significant grinding of the austenite grain and an increase in the reserve of viscosity. Another direction is microalloying with sorption-active elements that suppress the release of sulfides (Solano-Alvarez & Bhadeshia 2024, Podolskyi 2023) and carbonitrides along grain boundaries. Boron, under the condition of deep deoxidation with calcium, is not only a horophilic element, but also increases hardenability (Podolskyi 2023).

DOI: 10.1201/9781003518518-33

2 METHODOLOGY

2.1 *Production of test steel*

The production of experimental steel was carried out in laboratory conditions using the induction module ITPE-0.01/0.02 by remelting steel scrap with a carbon content of 0.72% with the addition of boron-containing ferroalloy of high purity in order to increase hardenability and as a horophilic element according to prior deep deoxidation with calcium-containing material. The introduction of deoxidizing and alloying materials into liquid steel was carried out by forcing them to be immersed in a protective metal container. As a result of steel production, an ingot was obtained with an actual weight of 9.2 kg, with a height of 120 mm and a diameter of 80 mm, with an actual chemical composition (Table 1).

Table 1. The actual chemical composition of the experimental steel, % mas.

C	Si	Mn	Al	V	B	Ca	N
0,9	0,39	0,89	0,01	0,001	0,0003	0,0006	0,012

2.2 *Hot plastic deformation*

Before hot plastic deformation, cutting of the experimental sample with dimensions of 70x70x80 mm was performed with a band saw with water cooling. The process of hot plastic deformation was carried out using laboratory equipment: a mine-type laboratory furnace and a universal testing hydraulic press CD-40/10. After heating the test sample to a temperature of 1250°C with subsequent exposure for 30 minutes, the test sample was placed on the working surface of the hydraulic press. With the gradual application of the load, the sample was deposited to a degree of deformation of 50%, which allows us to assess the effect of hot plastic deformation during the formation of the railway rail head in the process of the current manufacturing technology, and the sample was cooled in still air.

2.3 *Heat treatment*

Heat treatment was performed using a specialized SNOL-2.5/1200 laboratory furnace and included heating to 900±2°C with a holding time of 50 minutes. with subsequent controlled cooling rates of 0.52°C/s, 2.3°C/s, and 5.1°C/s. During tempering of test samples, after accelerated cooling from a temperature of 900±2°C, heating to a temperature of 200±4°C was applied with a 2-hour exposure time.

2.4 *Microstructural characterisation*

Metallographic studies of steels were carried out with the help of microscopes manufactured by Axiovert 200 M MAT models. The preparation of the grindings was carried out mechanically using specialized equipment for the preparation of metallographic grindings of the model "Labopol-5" with the use of the necessary diamond suspensions. Electrochemical polishing of samples was performed using glacial acetic acid (CH_3COOH) and chromic anhydride (CrO_3). The interlaminar distance was studied on a SEM-106 scanning electron microscope.

2.5 *JMatPRO simulations*

The calculation of the phase diagram was performed in the specialized software environment JMatPRO using thermodynamic databases with the back diffusion of the primary phase, calculated in steps of 1°C. The separation module from a solid solution was used to study the evolution of the formation and stabilization of structurally free cementite during heat treatment processes. The results of optical and electron microscopy were used to check the adequacy and adaptation of the simulation.

2.6 *Mechanical properties*

Tensile tests were performed on cylindrical samples with a diameter of the working zone of 5 mm, manufactured in accordance with DSTU ISO 6892-1:2019. The hardness control was carried out in accordance with the requirements of the DSTU EN ISO 6506-4:2019 standard. The determination of impact strength was performed in accordance with the requirements of DSTU ISO 148-1:2022 on test samples with a U-shaped cut with a radius of 2.5 mm.

3 RESULTS

As a result of researching the structure of test steel samples obtained in laboratory conditions after smelting and subsequent hot plastic deformation, it was established that the structure mainly consists of pearlite. Analysis of the structure after etching in sodium picrate did not reveal the presence of a cementite network along the boundaries of former austenite grains, which indicates the assimilation of boron in the structure with suppression of the formation of structurally free cementite. The results of the mechanical tests are given in Table 2.

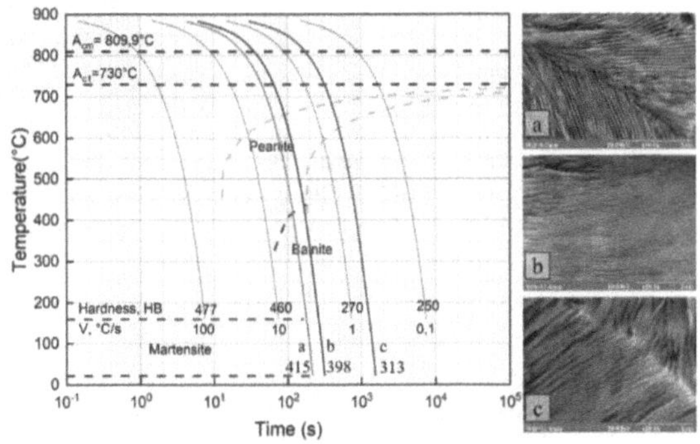

Figure 1. Calculated austenite decay kinetics of the test steel using JMatPro: a - pearlite structure after cooling at a rate of 0.52°C/s; b - pearlite structure after cooling at a rate of 2.3°C/s; c - pearlite structure after cooling at a rate of 5.1°C/s.

When analyzing the calculated thermokinetic diagram (Figure 1), obtained as a result of mathematical modeling using the JMatPRO software environment, it was the absence of structurally free cementite in the experimental steel with boron and calcium micro-alloying was confirmed. It is shown that the structure in the speed range of 0.52...5.3°C/s mainly consists of pearlite and a small amount of bainite - up to 2%. As a result of heat treatment carried out in laboratory conditions with the application of actual cooling rates, it was established that the structure mainly consists of pearlite of different dispersion: when steel is cooled at a rate of 0.52, 2.3, and 5.1°C/ with the interlaminar distance is 0.17, 0.11, 0.10 μm, respectively.

The research results showed that at a cooling rate of the experimental steel in the range of 2.3...5.1°C/s, the structure consists of highly dispersed pearlite with a hardness in the range of 398...415 HB. After cooling the test steel in accordance with the recommended range of cooling rates, tempering was carried out at a temperature of 200°C. The dependence of the influence of the cooling rate on the inter-plate distance and the hardness of the test steel was established (Figure 2). After a complete cycle of heat treatment, a test was conducted to determine the mechanical properties (Table 2).

Thus, it was established that the experimental chemical composition and heat treatment allow to achieve a complex of mechanical properties stiffness, which exceeds the requirements for steel provided by DSTU 4344:2004 and meets the requirements of DSTU EN 13674-1:2018 for railway rails.

Table 2. Mechanical properties of the test steel after heat treatment (5.1°C/s) according to the recommended regimes.

Process	UTS, MPa	YS, MPa	δ, %	Ψ, %	KCU, J/cm2	Hardness, HB
Hot plastic deformation	1076,4	604,5	9,5	12,6	9,3	280
Heat treatment	1295,5	816,4	9,8	11,4	17,25	368

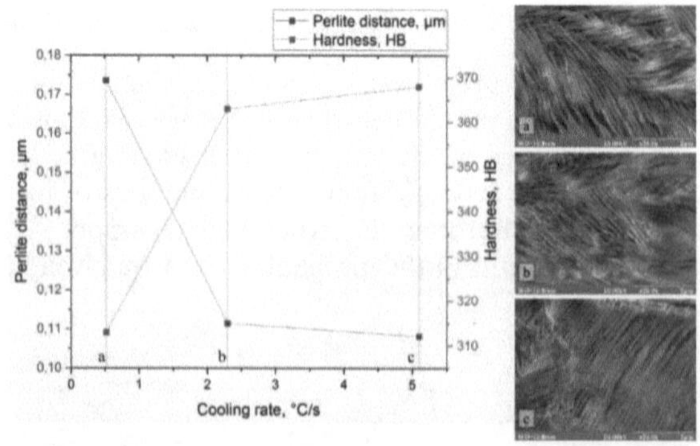

Figure 2. The influence of the cooling rate of the experimental steel on the inter-plate distance and hardness: a – cooling rate of 0.52°C/s; b - cooling rate 2.3°C/s; c - cooling rate 5.1°C/s.

4 CONCLUSIONS

1. Thermokinetic curves of phase transformation of steel with C=0.90%, Mn=0.89%, Si=0.39%, B=0.0003%, Ca=0.0006% were obtained. From the results of microstructural studies, it was established that at a cooling rate of up to 5.1°C/c, a pearlite structure of different dispersion is formed, the calculation method showed that at a cooling rate in the range of 5.3…12.8°C/c, the structure consists mainly of pearlite and martensite. At a cooling rate of 12.8°C/c and more, the structure consists mainly of martensite.
2. The obtained dependence of the dispersion of pearlite of the test steel with C=0.90%, Mn=0.89%, Si=0.39%, B=0.0003%, Ca=0.0006% on the cooling rate in the range of speeds, where the decay of austenite occurs according to the diffusion mechanism (up to 5.1°C/s). The regularity of the change in the hardness of the specified steel depending on the interlaminar distance of pearlite has been established.
3. The following chemical composition is recommended for experimental and industrial testing of railway rail manufacturing technology: C = 0.90%, Si = 0.39%, Mn = 0.89%, Ca = 0.0006%; B=0.0003% and recommended heat treatment parameters: accelerated cooling of the rail head at a rate of 5.1±0.2°C/s with subsequent tempering at 200°C.

REFERENCES

Zhou T. et al. 2023. A Novel Carbide-Free Bainitic Heavy-Haul Wheel /Steel with an Excellent Wear-Resistance under Rolling-Sliding Condition. *Metals* 13: 202.

Skoblo T.S. 2006. Struktura i svojstva zaevtektoidnoj stali posle termoobrabotki. Visnik HNTUSG im. P. Vasilenka 42: 8–11 [in Russian].

Poschitz I.N. & Kolesov V.A. 1998. Use of High-Nitrogen non magnetic steel for production of steel-aluminum conductor. *5-th Int. Conf. High Nitrogen Steels. Espoo*: 19.

Mujica Roncery L. et al. 2011. Nucleation and precipitation kinetics of $M_{23}C_6$ and M_2N in an Fe-Mn-Cr -C-N austenitic matrix and their relationship with the sensitization phenomenon. *Acta Materialia* 59: 6275–6286.

Boettinger W. J. et al. 2002. Phase-field simulation of solidification. *Annual Review of Materials Research* 32: 163–194.

Solano-Alvarez W. & Bhadeshia H.K.D.H. 2024. Steels for rails, *Progress in Materials Science* 146: 101313.

Podolskyi R.V. 2023. Development of chemical composition and modes of heat treatment of high-strength rails of pearlite class steels. PhD thesis., Dnipro.: Ukrainian State University of Science and Technologies.

Welding and Related Technologies – Krivtsun et al. (Eds)
© 2025 The Author(s), ISBN 978-1-032-85176-1

Physicomechanical properties of refined copper as a functional material for the working walls of crystallizers of continuous casting machines for billets

V.O. Tunyk & O.M. Smirnov
Physico-technological Institute of Metals and Alloys of the National Academy of Science of Ukraine, Kyiv, Ukraine

V.V. Dovbenko
LLC Production Enterprise Ukrkabel, Kyiv, Ukraine

A.Yu. Tunik
E.O. Paton Electric Welding Institute of the National Academy of Science of Ukraine, Kyiv, Ukraine

ABSTRACT: The primary functional component of continuous casting machines (CCM) for metals and alloys is the crystallizer, which ensures the formation of the billet's geometric shape and solid shell. The working walls of crystallizers for steel and copper casting are typically made of cathode copper M1 or high-purity alloyed copper alloys. The service life of a crystallizer largely depends on the correct choice of material for the crystallizer walls and the optimization of its design elements. This paper describes the technology for producing copper plates from secondary raw materials in a fire-refining furnace. The results of studying the physico-mechanical properties of plates made from fire-refined copper are presented. The possibility of using these plates as a multifunctional material for the working walls of crystallizers is substantiated. The conducted studies have shown that fire-refined copper is a promising functional material and can be considered a starting material for the production of plates for the working walls of rectangular crystallizers used in continuous casting of metals.

1 INTRODUCTION

The crystallizer in continuous casting machines for metals and alloys is the primary functional component responsible for forming the geometric shape and solid shell of the billet. The metal casting process involves intensive heat removal from the billet through the working walls of the crystallizer, which places rather high demands on the material (Smirnov et. al.2010). Typically, the working walls of crystallizers for steel and copper casting are made of cathode copper M1 or high-purity alloyed copper alloys (Bredikhin., et. al. 2006). However, there is no consensus on which alloying additions (Ag, Zr, P, Ni, Cr, etc.) and in what quantities should be preferred when selecting an alloy for the working walls of the crystallizer. The correct choice of material for the crystallizer walls, combined with the optimization of its design elements, can significantly increase the service life of the crystallizer and ensure high quality of the surface and subsurface layers of the continuously cast billet. In non-ferrous metal processing plants, high-purity copper is obtained by fire-refining, using various secondary copper raw materials as feedstock (Thomas, 2001), (Savenkov et. al. 2008).

DOI: 10.1201/9781003518518-34

2 OBJECTIVE AND TASKS OF THE RESEARCH

This paper describes the technology for producing copper plates from secondary raw materials using the fire- refining method. The aim of this work was to conduct a comprehensive analysis and evaluation of the possibility of using plates made of refined copper as working walls for crystallizers, taking into account its physico-mechanical properties as a multifunctional material. The research tasks included determining the electrical resistivity as an indirect indicator of thermal conductivity; the relative elongation of samples under tensile stress until failure as an indicator of plasticity; the number of revolutions in the twisting test of samples until failure as an indicator of thermal cycling resistance; the recrystallization start temperature as an indicator of thermal stability; the material's ability for strain hardening; and the material's resistance to oxidation (mass loss upon heating in an oxidizing atmosphere) to adjust the parameters of hot rolling.

3 EXPERIMENT

A FR-120 fire-refining furnace was used for the processing of secondary copper raw materials. The refining furnace is equipped with two burners, one of which operates on an air-gas mixture, while the second is an additional oxygen burner and is used to accelerate the melting process of scrap. Above the level of the melt mirror, there are usually five blowing tuyeres. In the rear wall of the furnace, there are blowers that, during melting, are submerged below the mirror by tilting the furnace and air or gas is supplied to them, depending on the course of a particular metallurgical process. In the process of copper refining, special fluxes are used, which, when melted, interact with the melt. The products of interaction, resulting from the chemical reaction, pass into the slag. After each addition of flux, the slag is removed from the surface of the melt through a slag window. The refined copper smelted in this way is cast into rectangular billets (thickness 200 mm) on semi-continuous casting machines (Izmailov, et. al. 2006).

The peculiarities of such casting include a relatively low billet withdrawal speed and a high intensity of heat removal from the molten metal, which, in combination with the high thermal conductivity of copper, creates rather complex operating conditions for the crystallizer walls. Moreover, the solidification process of a copper ingot and most of its alloys usually ends within the crystallizer, which has a length of 300 - 550 mm (Savenkov et. al. 2009). Ingots are cast up to 5 m in length and then cut into measured lengths of up to 1760 mm, determined by the width of the heating furnace. After heating to a temperature of 800-850 °C, the ingots are rolled on a DUO800 reversing mill into billets (slabs) of the required thickness (5-120 mm). The width of the slabs can reach 1000 mm, and the length 6000 mm. If necessary to obtain work-hardened metal, the slabs can be cold-rolled on a Quarto400 reversing mill from a thickness of 60 mm with a deformation degree of 30-50%.

4 DISCUSSIONS OF RESULTS

Experimental samples of plates for continuous casting molds (CCM) and semi-continuous casting molds (SCCM) were manufactured from copper obtained through fire refining of various secondary raw materials. Their performance characteristics were determined under production conditions. The macrostructure of ingots obtained from ingots cast from refined and pure copper are shown in Figure 1.

Figure 1 shows samples obtained from ingots cast from M1 grade copper and fire-refined copper. The microstructures of these samples differ significantly from each other. The M1 copper sample has a dendritic structure typical of cast metal. Columnar dendrites are formed from the center and extend through the entire ingot to the edge. Such crystals have a thickness of 1.0 to 5.0 mm and a length of up to 50 mm. The refined copper ingot sample has an equiaxed grain structure. In the central part, the grain size is 5...10 mm, and closer to the edge it decreases to 1...3 mm. Such a significant difference in structure will obviously affect the properties of the samples under study.

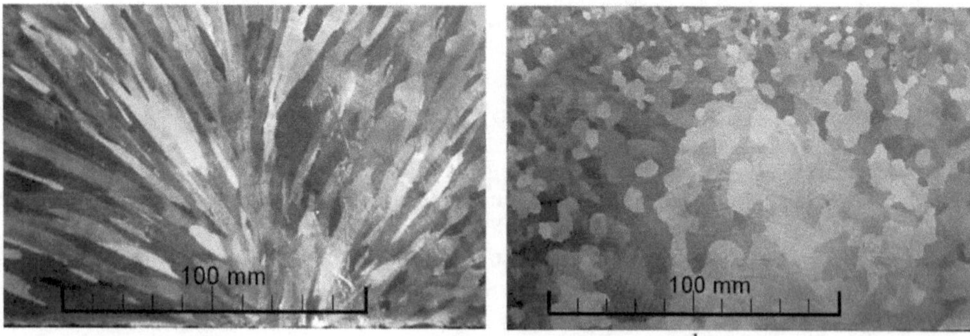

a b

Figure 1. Macrostructure of ingots cast from pure (a) and refined (b) copper; magnification x10.

When comparing the chemical composition of copper samples of grade M1 and refined copper of grade CuFRTP, the first significant difference is in the content of silver, tin, nickel, and phosphorus (Smirnov, et. al. 2024). It's noteworthy that these elements are commonly used for microalloying copper to improve its thermal stability (Table 1). Based on this, we can hypothesize that the recrystallization start temperature of products made from refined copper will be higher than that of cathode copper, which will enhance the operational durability of the crystallizer.

Table 1. Chemical composition of copper samples.

Material	%	Element content, ppm										
	Cu	Fe	Pb	Sn	Sb	As	Ni	Zn	Ag	Bi	P	S
Cathode copper M1	99.993	17	8	2	3	1	4	3	10	2	5	8
Fire-refined copper	99.957	2	42	67	15	4	109	2	150	4	18	6

To assess the influence of chemical composition on the thermal conductivity of copper, electrical resistivity was used as an indicator. The resistivity was measured on wire rod samples rolled on the same mill from both cathode and refined copper. The electrical conductivity of the material was determined by the eddy current method using a "SIGMATEST 2.069" device at a current frequency of 60 kHz. The conversion of conductivity values into thermal conductivity values can be carried out, for example, using the well-known Wiedemann-Franz law (Savenkov et. al. 2009); (Gridin et. al. 2009).

Results of the statistical analysis of the data (420 samples) show that the average electrical resistivity of wire rod made from refined copper is $0.017131 \cdot 10^{-6}$ Ohm·m, while the resistivity of wire rod made from cathode copper is $0.017101 \cdot 10^{-6}$ Ohm·m.

Moreover, the influence of the main accompanying chemical elements on the electrical conductivity (γ) of the wire rod made from refined copper can be represented by the following correlation dependence (chemical element content in %):

$$\gamma = 0{,}01705 + 0{,}00233\ Sn + 0{,}006355\ Ni - 0{,}00129\ Ag \quad (1)$$

This dependence was obtained based on measurements of electrical resistivity in 600 samples of 8 mm diameter copper wire rod (samples were taken from coils weighing 3-3.5 tons). In total, data from 21 melts, each weighing 110 tons, were included in the analysis. All melts were cast using the same technology, but there were inevitable deviations in the technological parameters within the established tolerances. Overall, the obtained data indicate that if the working walls of the crystallizer are made of refined copper, the heat removal intensity will remain at the same level as for cathode copper.

To analyze the ductility of refined copper, wire rod samples were also used, which were tested for tensile strength, after which the value of relative elongation was determined. For the studied samples of cathode wire rod, the impurity content was 0.007%, and for refined copper - 0.043% (i.e., 6 times higher). It was found that the value of relative elongation of samples made of cathode copper is within the range of 44-46% (90% of samples) and higher, while for samples made of refined copper it is slightly lower: 40-42% (30% of samples), 42-44% (50% of samples), and 44-46% (20% of samples). However, all these values exceed the minimum allowable values for soft-annealed products according to ASTM B152.

The resistance of the material to thermal cycling was evaluated on 8 mm diameter wire rod samples by comparing the number of twists to failure. As shown in Table 2, cathode copper can withstand a slightly larger number of twists than refined copper. This is apparently due to the presence of tin and lead impurities in refined copper, which are mainly located at grain boundaries. However, the difference in the measured values is not so critical as to prevent the use of refined copper as a functional material.

Table 2. Comparison of the twist value of samples for refined and cathode copper.

Material	Number of revolutions before failure				
	50-60	60-62	62-64	64-66	66-70
Cathode copper, %	11	3	4	33	49
Fire-refined copper, %	14	36	35	11	11

The recrystallization start temperature of the compared copper grades was determined by annealing rolled samples with a preliminary deformation degree of 95% and a thickness of 0.4 mm in a laboratory muffle furnace at various temperatures. The annealing duration was 60 minutes (Figure 2).

It was found that the recrystallization start temperature of refined copper is approximately 45° C higher than that of cathode copper. This can be attributed to the presence of certain amounts of silver, nickel, and tin in refined copper, which are inevitably present in the melt as accompanying elements. It is known that silver is practically not removed from the melt by fire refining methods, and the content of tin and nickel has a certain lower limit, determined by the flux operation.

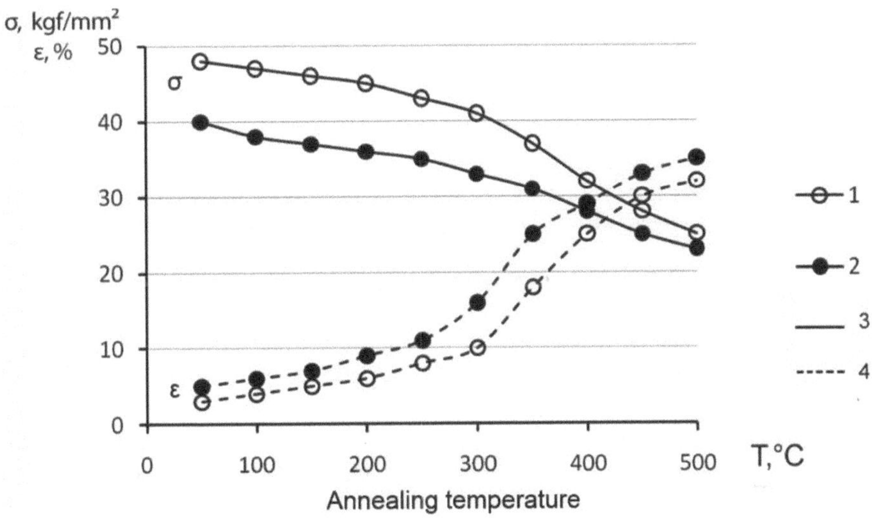

Figure 2. Softening curves of samples: (1) refined copper; (2) cathode copper; (3) tensile strength, kgf/mm^2; (4) relative elongation, %.

The ability of refined copper to undergo work hardening was evaluated by constructing strain hardening curves during tensile tests. Tensile tests were performed on an IR 5040 tensile testing machine. The allowable error in conducting tensile tests was +/-1%. 18 mm diameter wire rod was used as the initial samples. As the studies showed, for samples made of refined copper, the ultimate tensile strength is higher in the entire studied range. For example, at a deformation degree of about 50%, the ultimate tensile strength for samples made of refined copper is 30 MPa higher than for samples made of cathode copper (almost 10%). In general, the increase in ultimate tensile strength for refined copper should be considered a positive factor in terms of increasing the strength and durability of parts or products operating under complex thermal and mechanical loads. Such products include the working walls of crystallizers, which are subject to warping.

The tendency of refined copper to oxidize during heating is a very important technological parameter, since the working surface of the plates is heated during casting and comes into contact with the atmosphere located in the gap between the billet and the crystallizer walls (Shinkarenko et. al. 2009). A comparison of the oxidation tendency of refined and cathode copper was conducted by simultaneously heating several samples in a laboratory muffle furnace. These samples were previously weighed on analytical balances. After a specified thermal treatment time, a pair of samples was removed from the furnace and immediately immersed in water. As a rule, most of the formed scale, due to rapid cooling, was destroyed and separated from the samples. The remaining scale was removed from the surface of the samples by chemical etching. After that, they were reweighed. The magnitude of the weight loss of the samples was used to evaluate the material's tendency to oxidize. Generalizing the obtained data, it should be noted that refined copper oxidizes more intensively when heated than cathode copper, on average by 18-20%. In our opinion, this should be attributed to the higher oxygen content in fire-refined copper.

5 CONCLUSIONS

A comprehensive analysis of the physicomechanical properties of cathode (M1) and refined (CuFRTP) copper has shown the following:

- In the case of manufacturing working walls of the crystallizer from refined copper, the heat removal intensity will remain at the same level as for cathode copper.
- For crystallizers made of fire-refined copper, the softening temperature is 410-440°C, and the beginning of recrystallization is 370-400°C. These indicators are not inferior to similar indicators of materials for crystallizers made of cathode copper.
- It was established that for refined copper, the temperature of the beginning of recrystallization is approximately 45°C higher than for cathode copper. This should ensure increased operational durability of the crystallizer.
- A study of plasticity indicators showed that the value of relative elongation of samples made of cathode copper is within ≤45±1% and higher, while for samples made of refined copper it is slightly lower: 41±1%. However, these values exceed the minimum allowable values for rolled products according to ASTM B152.
- Tensile tests showed that for samples made of refined copper, the tensile strength is ~10% higher throughout the entire studied range.
- Comparing the tendencies of refined and cathode copper to oxidation, it should be noted that refined copper oxidizes more intensively (~ by 20%) when heated than cathode copper. This should be attributed to the higher oxygen content in fire-refined copper.

The conducted studies of the structure and physicomechanical properties have shown that fire-refined copper is a promising multifunctional material and can be considered as a starting material for the manufacture of plates for the working walls of rectangular crystallizers used in continuous casting of metals.

REFERENCES

Bredikhin V.N., Manyak N.A. & Kaftanenko A.Ya. (2006). Secondary copper: monograph. Donetsk: DonNTU, 416 p.

Gridin S.V. et al. (2009). Study of shrinkage phenomena during the formation of copper ingots. *Journal.* Metal Science and Heat Treatment 2009, (2): 35–39.

Izmailov V.A., Friedlyanskyi, V.F., Goloveshko, A.I. & Suvorov (1992). New designs of crystallizers for continuous casting of copper and copper alloys. *Journal.* Color metals. 1992. (7): 56–58.

Savenkov Yu.D. et al. (2008) Rafinirovannaya med' Ukrainy. Dnepropetrovsk: ART-PRESS, 2008. 176 p.

Savenkov Yu.D. et al. (2009) Study of the process of rectangular ingot formation during semi-continuous copper casting. Scientific Works of DonNTU. Series: Metallurgy. 2009. (11): 253–261

Smirnov O.M., Kuberskyy S.V. & Shtepan Ye.V.. (2010) Bezperervne rozlyvannya stali. Textbook. Alchevsk: DonDTU. 2010: 520 p. ISBN 978-966-310-250-4

Smirnov O.M., Narivskiy A.V., Ukhin V.Ye. & Tunyk V.O. (2024) The effect of phosphorus content in copper alloys used for crystallisers of continuous casting machines *Journal.* Met. lit'e Ukr., vol. 32, 2024. No2 (337): 31–34

Shinkarenko P.S., Savenkov Yu.D. & Kvachenyuk N.Ye. (2007). Kompleks ognevogo rafinirovaniya medi na OAO «Artemovskiy zavod po obrabotke tsvetnykh metallov» *Journal.* Metall. 2007. (3): 60–62.

Thomas B.G. (2001) Continuous casting: Complex Models. The Encyclopedia of Materials: Science and Technology. – Oxford: Elsevier Science Ltd. (Vol. 2): 1599–1609.

Welding and Related Technologies – Krivtsun et al. (Eds)
© 2025 The Author(s), ISBN 978-1-032-85176-1
Open Access: www.taylorfrancis.com, CC BY-NC-ND 4.0 license

Effect of heat treatment on the structure of the Fe-28Mn-12Al-0.9Si-1.35C alloy, additionally doped by Ni 3.3, Cr 1, V 0.15 (wt. %)

O.M. Smirnov, M.M. Voron, A. Tymoshenko, A. Semenko & Yu. Skorobagatko
Physico-Technological Institute of Metals and Alloys of the National Academy of Sciences of Ukraine, Kyiv, Ukraine

S.L. Schwab
E.O. Paton Electric Welding Institute of the National Academy of Sciences of Ukraine, Kyiv, Ukraine

ABSTRACT: Microstructure of the novel high-strength steel Fe-28Mn-12Al-0.9Si-1.35C after additional alloying and thermal treatment was investigated. Nickel and chrome in quantity of 3,3 and 1 wt. % respectively were added to the basic alloy to obtain less primary ferrite and to ensure austenite alloying rate. Vanadium was added to refine structure and to elevate mechanical properties. All three alloys were investigated in four states: as-cast, annealed, quenched and aged after quenching.

1 INTRODUCTION

High-strength lightweight steels have a large amount of chemical compositions, and there are some of them, that are able to change mechanical properties in very wide range due to thermal or thermomechanical treatment. They usually contain (wt. %) 20-35 Mn, 6-12 Al and 0.6-1.5 C. Proportion of main alloying elements determine steels' ability to change mechanical properties in a very wide range (Hu et al. 2022). According to (Wietbrock et al. 2011), high manganese concentration is beneficial for dissolution of alloying elements, dynamic recrystallization and formability. Average carbon content in high-Mn steels is often between 0.6-1.2 wt. %. High aluminum content determines higher percent of carbon, because of k-carbides formation. In fact, effective Al content lays in a range 5-12 wt. % and causes alloys significant density decrease by 10-15 %.

Addition of alloying allows to enhance phase-structure and mechanical characteristics. For example, is 0.2-1 wt. % silicon strengthens steel and prone martensite formation during deformation. Chromium and nitrogen can improve yield strength and low-temperature toughness. Nickel elevates strength and ductility and causes desirable changes in carbides formation (Rahnama et al. 2017). Niobium is good for recrystallization inhibition, but not effective for grain refinement (Hu et al. 2022, Zhao et al. 2023), while vanadium is a very profitable element for grain refining and significant carbide strengthening. Moreover, it also strengthens grain boundaries (Gwon et al. 2017.).

Steels with austenite matrix and small amount of ferrite in initial state, often demonstrate high level of strength and sufficient plasticity after thermal treatment. So, in this work, basic steel composition was chosen in such way to dissolute significant amount of aluminum, silicon and carbon in austenite matrix and to obtain minimum ferrite, as a determinative sign of optimum Al and Si saturation and to have ability for its dissolution due austenisation and quenching. In this way, better precipitation hardening may be occurred. After this, addition alloying with Ni and Cr was provided, as well as V addition for structure refining.

DOI: 10.1201/9781003518518-35

2 MATERIALS AND METHODS

As-cast steel ingots were obtained by induction melting and sand casting. Significant amount of secondary materials and ferroalloys were used. Basic alloy contained only main elements – Fe, Mn, Al, Si and C. Second alloy contained nickel and chromium, which were added to strengthen alloy and to enhance austenite existing area. Third alloy contained vanadium as a grain refining modificatory and element, that makes precipitation hardening more effective due to formation of more fine complex carbides. Chemical composition of alloys is shown in Table 1.

Table 1. Chemical composition of experimental alloys.

	Chemical composition, wt. %, Fe – balance						
Alloy	Mn	Al	C	Si	Ni	Cr	V
1	28.45	12.70	1.37	0.80	-	-	-
2	28.50	12.30	1.30	0.90	3.35	1.00	-
3	28.10	11.90	1.38	0.90	3.40	1.00	0.13

Specimens were mechanically cut from ingots, after what thermal treatment has been occurred (annealing, quenching and hardening). Then they were prepared for metallographic and micro-durometric testing by grinding, polishing and etching with 3% ethyl solution of nitric acid.

Microstructure changes were evaluated and compared in different states by using optical microscope iScope IS.1053-PLMi. After this, microhardness tests were provided on LHVS-1000Z testing machine. They helped to determine influence of thermal treatment indirectly by evaluating some changes in phases like solid solution alloying rate and strengthening rate.

3 RESULTS AND DISCUSSION

Basic alloy in as-cast state demonstrates dendritic structure of large austenite grains without visible ferrite parts and large grain boundaries. Structure becomes homogeneous after annealing and obviously consists of austenite and some small particles, that may be as ferrite as carbides. Quenching caused fixation of fully austenite structure and coarsening of grain boundaries. Small symmetrical inclusion formed in austenite grains after aging. They may correspond to places of former dendritic branches boundaries. Thermal treatment helped to discover large grain boundaries of macro-grains, what may be undesirable.

Microstructure of as-cast samples refines after nickel and chromium addition. Thin primary formed ferrite dendrites were found, despite significant nickel concentration. It means, that cooperation of silicon and chromium is strong enough to form ferrite in conditions of enhanced austenite stabilization. Ferrite dendrites do not grow after annealing, but large columnar grains become coarser. Quenching provides ferrite dissolution, but primary dendritic structure retains, possibly because of grain boundaries coarsening. Age hardening provides formation of thin ferrite-like particles that have different orientation and dimensions. They are probably deposited on the austenite dendrites grain boundaries instead of dendrite centers, like in as-cast state.

Alloy that was modified with vanadium demonstrates combination of structural features of previous two alloys. It has relatively fine dendritic structure with small globular ferrite particles inside. After annealing, ferrite particles are deposited at the grain boundaries of austenite. After quenching, third alloy is similar to second with only difference, that ferrite phase is not fully dissolute in austenite. Age hardening also provides ferrite precipitation on grain boundaries, but as soon as vanadium is also an alpha-stabilization element, these particles are bigger, then previous.

It both cases, it is obvious, that for such concentration of ferrite stabilization elements, chosen austenization temperature is almost sufficient. It is also clear, that additional alloying with nickel and chromium allows to change structure-phase state in wide range by using heat treatment. It also allows to lower aging temperatures under 450 °C, where ferrite is already formed.

Microstructures of obtained samples are shown in Figure 1.

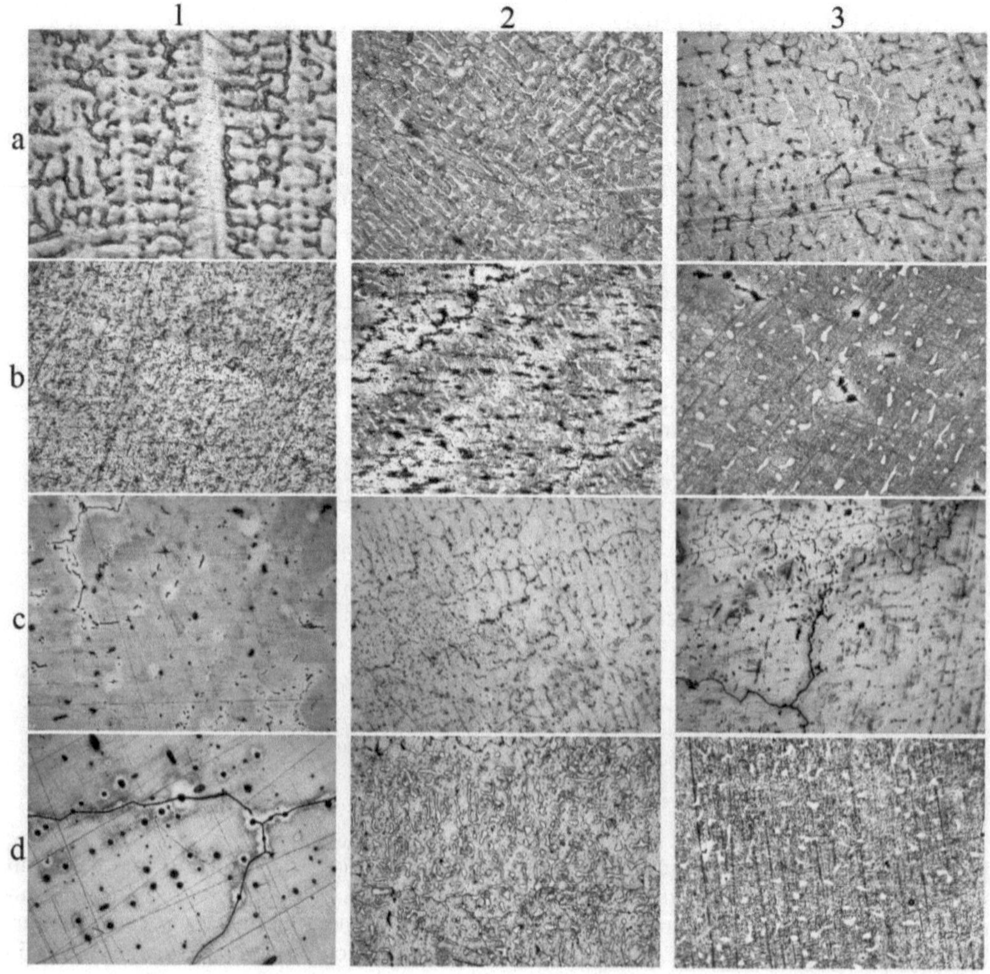

Figure 1. Samples microstructure: a – as-cast; b – annealing 1100 °C 4 hrs; c – quenching (1100 °C 4 hrs, rapid cooling in water); d – age hardening (450 °C 4 hrs after quenching). Magnification X200.

Microhardness test (Table 2) shows, that all three alloys in as-cast state has comparably similar properties of ferrite and austenite. Microhardness after annealing decreases for both phases, except ferrite in third sample with vanadium microalloying, where this property remains almost the same and austenite microhardness decreases less, then in samples 1 and 2.

Quenching of all the alloys provides almost the same austenite microhardness, while further age hardening leads to formation of sharply different results. In basic alloy and sample with Ni and Cr austenite hardness rises to same value, while in V-doped sample, it decreases, becoming lower by 20 % then other two alloys. Compositions with Ni and Cr have ferrite particles with high microhardness for undoped alloy and almost 10 % lower for alloy with vanadium, while its austenite microhardness was 20 % lower.

4 CONCLUSIONS

Basic alloy has predictable structure and phase changes after thermal treatment and has proportional changes in phases microhardness, what may lead to good deformability and mechanical properties. Alloy with Ni and Cr addition has significantly refined structure and much lower uniformity of properties in as-cast and annealed states. After quenching and age

Table 2. Microhardness of the experimental alloys in as-cast state and after thermal treatment (according to Figure 1).

	Vickers micro hardness of phases											
	1				2				3			
	Ferrite		Austenite		Ferrite		Austenite		Ferrite		Austenite	
Sample	Hv^{50}	HRC	Hv^{50}	HRC	Hv^{50}	HRC	Hv^{50}	HRC	Hv^{50}	HRC	Hv^{50}	HRC
a	622	57	577	54	606	56	538	52	614	56	531	51
b	514	50	480	48	514	50	382	39	621	57	488	48
c	-	-	551	53	551	53	514	50	-	-	538	52
d	-	-	606	56	637	57	606	56	550	52	480	48

hardening this alloy may obtain highest strength due to highest microhardness increase, what potentially points to highest strength. Chromium content ensures ferrite formation, so heating temperature should be raised for more effective thermal treatment. Vanadium addition provides more ferrite stabilization and leads to lower microhardness mismatch in as-cast state, but worst after age hardening. That may cause highest plasticity after thermal treatment.

ACKNOWLEDGMENTS

This research has been funded by National Research Foundation of Ukraine, competition "Science to Strengthen Defense Capabilities of Ukraine" (Project No. 2023.04/0021), "Production of armor elements for military equipment from ultralightweight steels using secondary raw materials". Chemical composition, optic metallography and micro hardness studies were conducted at the Center of Scientific Equipment Collective Use "Materials Science of Refractory Compounds and Composites" (National Technical University of Ukraine "Igor Sikorsky Kyiv Polytechnic Institute"), staff of which is gratefully acknowledged.

REFERENCES

Hu, S. et al. 2022. Fe–Mn–C–Al Low-Density Steel for Structural Materials: A Review of Alloying, Heat Treatment, Microstructure, and Mechanical Properties. *Steel research international.* 93: 2200191.

Wietbrock, B. et al. 2011. Effect of temperature, strain rate, manganese and carbon content on flow behavior of three ternary Fe-Mn-C (Fe-Mn23-C0.3, Fe-Mn23-C0.6, Fe-Mn28-C0.3) high-manganese steels. *Steel research international.* 82: 63–69.

Rahnama, A. et al. 2017. Effect of Ni alloying on the microstructural evolution and mechanical properties of two duplex light-weight steels during different annealing temperatures: Experiment and phase-field simulation. *Acta Materialia.* 132 (15): 627–643.

Zhao, T. et al. 2023. Influence of Thermo-Mechanical Process and Nb-V Microalloying on Microstructure and Mechanical Properties of Fe–Mn–Al–C Austenitic Steel. *Coatings.* 13(9): 1513.

Gwon, H. et al. 2017. The effect of vanadium micro-alloying on the microstructure and the tensile behavior of TWIP steel. *Mater. Sci. Eng. A.* 696: 416–428.

Session 8. Strength and stress-strain states of welded joints and structures

Welding and Related Technologies – Krivtsun et al. (Eds)

Scientific principles of magnetic pulse treatment of welded joints in the process of fusion welding

L.M. Lobanov, M.O. Pashchyn & O.L. Mikhodui
E.O.Paton Electric welding institute of the National Academy of Sciences of Ukraine, Kyiv, Ukraine

O.M. Karlov, I.P. Kondratenko, R.S. Kryshchuk & V.V. Chopyk
Institute of Electrodynamics of the National Academy of Sciences of Ukraine, Kyiv, Ukraine

P.R. Ustymenko & Yu.M. Sydorenko
National Technical University of Ukraine "Igor Sikorsky Kyiv Polytechnic Institute", Kyiv, Ukraine

ABSTRACT: treatment of welded joints with a pulsed electromagnetic field (TPEMF) reduces the level of residual welding stresses. TPEMF during welding increases the efficiency of the weld-ing process (compared to TPEMF after welding) and makes it easier to implement. On the basis of mathematical modeling and experimental studies of magnetic pulse processes, an automated complex for TIG welding has been developed that is compatible with the TPEMF of the weld metal under the conditions of a thermo-deformation welding cycle.

1 INTRODUCTION

The development of industry initiates the research of advanced technologies for improving the operational properties of welded structures. Treatment of welded joints with a pulsed electromagnetic field (TPEMF) is promising in this regard (Lobanov 2018, Batyigin 2003, Turenko 2009). Based on the results (Batyigin 2003, Turenko 2009), it should be noted that the action of the pulsed electromagnetic field (PEMF) is an effective tool for influencing the stress states of welded joints. At the same time, the effectiveness of the impact increases when it is used in welding conditions (Lobanov 2022). On the basis of TPEMF, effective technologies for controlling the stress-strain state of welded joints made of non-ferromagnetic metal materials, which include aluminum alloys used in the aircraft and shipbuilding industries, can be developed.

The aim of this work is to develop the scientific basis for the TPEMF application in the process of non-ferromagnetic materials welding using aluminum alloy AMg6.

2 RESEARCH METHODOLOGY

Increasing the efficiency of TPEMF in the welding process, in comparison with processing at room temperature, based on the principles: increasing the productivity of the welding process due to the transition from sequential to simultaneous main technological operations; possibility of automation; increasing processing efficiency to reduce residual welding stresses according to the mechanism (Lobanov 2022); dispersion of the weld metal structure during magneto-dynamic actions at elevated temperatures; the use of welding and assembly equipment elements (as screens) to increase the value of the electrodynamic pressure force P. As it known, eddy currents are excited when flowing along the conductors of an electric current pulse (ECP) inductor in an approximate electrically conductive medium. As proved in (Lobanov 2021, Vasetsky 2017), as a result of the interaction of the induced currents with the EMF, which excited these currents, an electrodynamic force P arises, the normal component of which exerts an active load

DOI: 10.1201/9781003518518-36

on the area and - changes the stress state of the material being processed. Provided that the density j of the induced electric current reaches the value of $j \geq 1$ kA/mm2 in the processed metal, conditions are created for the realization of the electroplasticity effect (EPE) (Lobanov 2022, Lobanov 2021, Vasetsky 2017). This contributes to the intensification of the plastic deformation of the material and, as a result, the relaxation of its residual stress state. A flat inductor with an outer diam. of 95 mm and an inductance of L = 120 μH was used to implement the TPEMF (Lobanov 2021). A power source based on a capacitor system with a total capacity of C = 5140 μF, a charging voltage of U up to 800V, and a stored energy of Es ~ 1.6 kJ was used to generate EMF. To evaluate the efficiency of the TPEMF, samples with a weld seam in the form of a disk with thickness δ =1.0 mm and a diameter Ds = 90 mm, made of AMg6 were used. Annular welded joints with full penetration were performed by the TIG method along the line of a circle with a diameter of Dw=45 mm. The method of electronic speckle interferometry (Lobanov 2021, Lobanov 2006, Wohlfahrt 2012, Wohlfahrt 2009) was used to evaluate the σ_x tangential component of the stress state of welded joints at points on the ring weld line.

3 RESEARCH RESULTS

TPEMF of samples with a thickness of δ =1.0 mm and assemblies of a sample of δ =1.0 mm with a conductive screen made of AMg6, also in the form of a disk of δ =5.0mm and 90mm in diameter, were performed. TPEMF was performed at a charge voltage of U =800V. The force P will increase with increasing thickness, as it defined as an integral value in a certain volume, which is confirmed by the data in Figure 1 and the results of (Lobanov 2021). When δ increases to 6.0mm (use a screen), P values increase more than twice as compared to TPEMF without a screen of samples δ =1.0 mm. For TPEMF of thin samples, a way is proposed in the form of installing additional layers of a related material, for which the equivalent thickness will be optimal from the point of view of achieving the greatest value of the electromagnetic pressure force P. The proposed way is similar to the application of metal screens in the technology of electromagnetic molding, such as dis-pensing or crimping (Bilyi 1977). Figure 2 shows the distributions of residual stresses σ_x be-fore TPEMF, as well as σ_x after TPEMF of the sample without and with the use of a screen, which confirm the effectiveness of shielding with TPEMF.

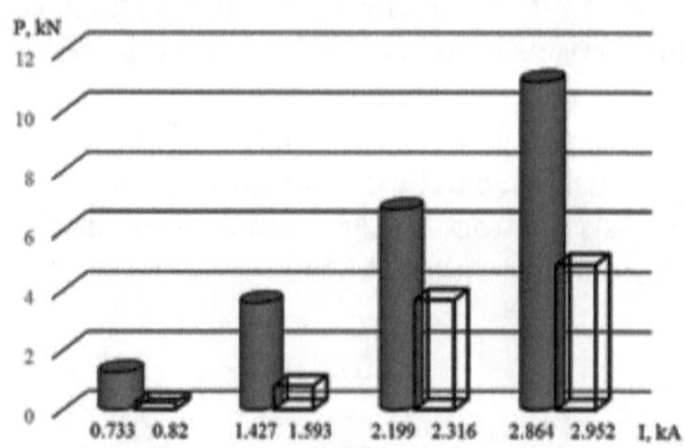

Figure 1. The influence of the amplitude values of ECP - I on the pressure force P during the TPEMF of samples of ring welded joints δ = 1.0 mm from the AMg6 alloy, where: □ – TPEMF without a screen, ■ – TPEMF with a screen.

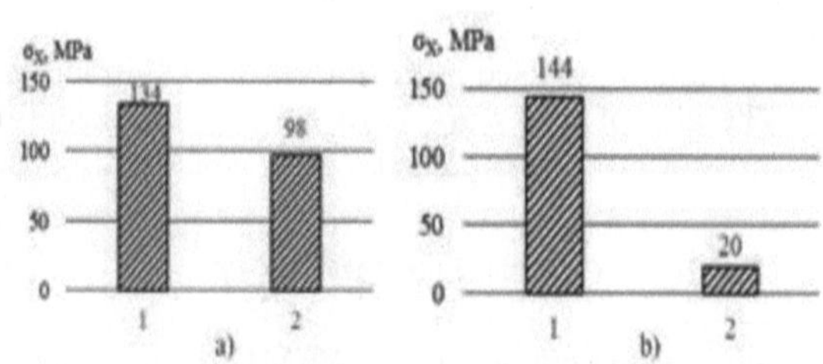

Figure 2. Influence of TPEMF of the weld without application (a) and with application (b) of the screen on the residual stresses σ_X in the metal of the AMg6 alloy welded joint, where: 1 - before TPEMF, 2 - after TPEMF.

Considering the above, it should be noted that TPEMF can be an effective means of managing the residual stress states of welded structures made of aluminum alloys. The results of metallographic studies of samples before and after TPEMF confirmed the positive effect of processing on the evolution of the weld metal structure (Figure 3). In general, the microstructure is a light matrix - an α-solid solution with the release of an excess phase in the form of a grid along the boundaries of the crystals. Without TPEMF, the crystals are elongated along the direction of their

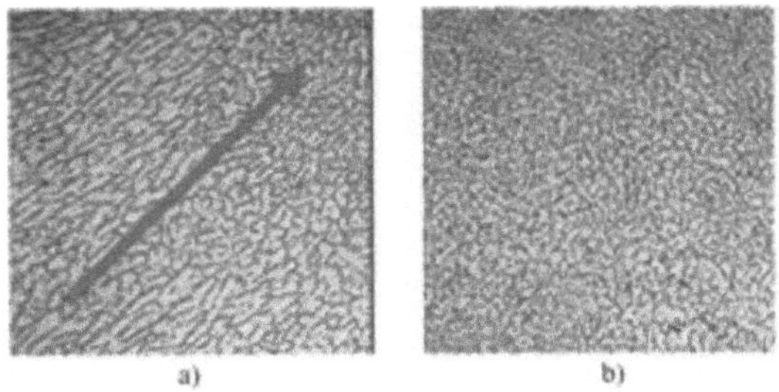

Figure 3. Microstructure (x200) of the weld metal of the AMg6 alloy: a) in the initial state (before TPEMF), where the arrow indicates the direction of crystal growth; b) after TPEMF.

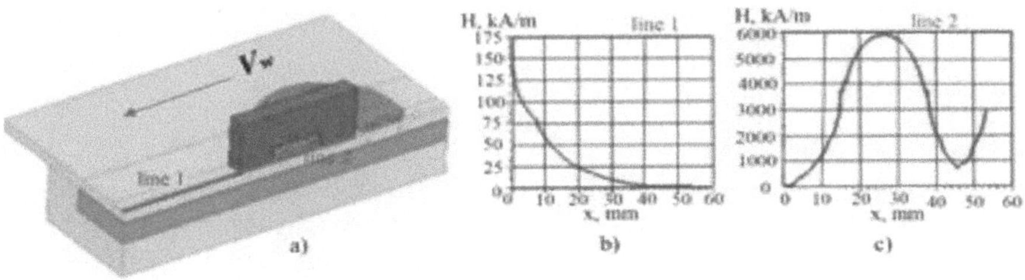

Figure 4. Mathematical modeling of the electrophysical parameters of the TPEMF (welded seam) in the welding process, where: a) the diagram of the positioning of the U-shaped inductor of the PEMF along the welding direction Vw (explanation in the text); b) distribution of H of PEMF voltage along line 1; c) distribution of the PEMF intensity H along line 2.

growth - normal to the fusion line of the weld (arrow in Figure 3a). After TPEMF with shielding, there is an equiaxial orientation of the grains (Figure 3b), which is accompanied by their dispersion. The obtained results can be explained by the dynamic pressure of the *P* force on the grain boundaries in the TPEMF process, which is accompanied by local Joule heating of the grains (Stryzhalo 2008). On the basis of the results (Rashchepkin 2019), mathematical modeling of the electrophysical parameters of the TPEMF of the weld seam at the welding process was carried out. The simulation results (Figure 4) should contribute to the optimization of the TPEMF parameters for the relaxation of the residual stresses of thin-sheet welded joints. Figure 4a shows the positioning scheme of the U-shaped inductor for TPEMF along the welding direction Vw (arrow), where line 1 is the surface of the weld in front of the inductor, line 2 - in the zone of direct action of the PEMF on the weld metal. Figure 4b shows the distribution of the magnetic intensity H of the PEMF, which allows determining the optimal position of the inductor relative to the welding torch, which is determined by the maximum value of H along line 1 (the beginning of the line on the side surface of the U-shaped inductor). It follows from Figure 4b that at a distance of 40 mm from the edge of the magnetic conductor (x = 0), the value of H is close to zero. The diagram in Figure 4b allows you to determine the distance between the inductor and the welding torch, at which the influence of PEMF on the welding process is minimal. Figure 4,c shows the distribution of H of TPEMF along line 2 (Figure 4, a). Figure 5 shows the results of modeling the time distributions of the electrophysical parameters of the TPEMF of the weld. In Figure 5,a section 1 with dimensions of 3.5x5 mm (corresponding to the rear front of the welding bath) is highlighted. For section 1 the average current density J (Figure 5,c) and the average value of the magnetic pressure P (Figure 5,d) are determined during the action of ECP - I (Figure 5,b) on the solidified weld seam, which contributes to the relaxation of welding stresses. At a time of about 50 μs, the average current density J on the metal surface is about 2200 A/mm2 (Figure 5,c), and the pressure P reaches its maximum value at 150 μs - 23 MPa (Figure 5,d). This should ensure the relaxation of welding stresses at the stage of their formation during solidification of the weld. Figure 5,e shows the distribution of the voltage H of the PEMF in the cross section of the inductor for the moment of the largest values of I and H of the PEMF, which ensure stress relaxation.

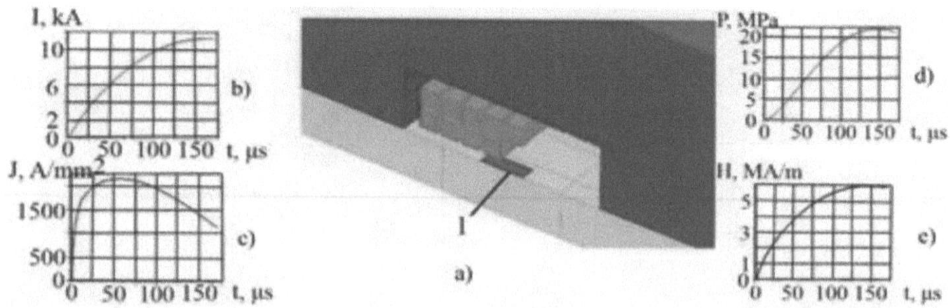

Figure 5. Results of simulation of time distributions of electrophysical parameters of TPEMF weld (explanation in the text).

4 CONCLUSIONS

On the basis of research results, an automated complex was developed for TIG - welding in conditions of TPEMF of solidifying weld metal (Figure 6), where a forming substrate -1 of non-ferromagnetic material acts as a screen (which enhances the effect of the force P) and the distance of the PEMF between the inductor 2 and the TIG torch -3 ensures the maximum efficiency of the PEMF during the welding process. In the near future, experimental studies of the effectiveness of the TPEMF weld technology in the welding process will be conducted using the complex (Figure6).

Figure 6. Automated complex for automatic welding in the conditions of the accompanying TPEMF of the weld, where 1 is the forming substrate, acting as a screen, 2 – inductor, 3 – burner, L $_{TPEMF}$ – distance between burner 2 and inductor 3.

REFERENCES

Batyigin Yu.V. (ed.), Lavinskyi V.I., Khimenko L.T. 2003. Pulsed magnetic fields for advanced technologies. Ukraine. Kharkiv. MOST-Tornado.

Bilyi I.V., Fertik S.M., Khimenko L.T. 1977. Handbook of magnetic pulse metal processing. Ukraine. Kharkiv. Vyscha shkola.

Lobanov, L., Kondratenko, I., Zhiltsov, A. at al. 2018. Development of Post-weld Electrodynamic Treatment Using Electric Current Pulses for Control of Stress-Strain States and Improvement of Life of Welded Structures. *Materials Performance and Characterization*, Volume 7, №. 4: 941–955.

Lobanov L.M., Pashchyn M.O; Mikhodui O.L at al. 2021. Pulsed Electromagnetic Field Effect on Residual Stresses and Strains of Welded Joints of AMg6 Aluminum Alloy. *Strength of Materials* Vol.53, Is.6: 834–841.

Lobanov L.M., Pashchyn M.O.; Mikhodui O.L. at al. 2022. Stress-Strain State of Welded Joints of AMg6 Alloy after Electrodynamic Treatment During Welding. *Strength of Materials*, Vol. 54, Is.6: 983–996.

Lobanov L.M., Pivtorak V.A., Savitsky V.V. and Tkachuk G.I. 2006. Procedure for determination of residual stresses in welded joints and structural elements using electron speckle-interferometry. *The Paton Welding Journal*. №1: 24–29.

Rashchepkin A.P., Kondratenko I.P., Karlov A.N. and Kryshchuk R.S. 2019. Electromagnetic field of w-shaped inductor for magnetic-pulse processing of materials. *Tekhnichna Electrodynamika*. № 6: 5–12.

Stryzhalo V.A., Novohrudskyi L.S., Vorobiov E.V. 2008. Strength of materials at cryogenic temperatures with electromagnetic fields effects. Ukraine. Kyiv. IPP.

Turenko A.N., Batyigin Yu.V., Hnatov A.V. 2009. Pulsed magnetic fields for progressive technologies.- V.3: Theory and experiment of attraction of thin-walled metals by pulsed magnetic fields: Monograph. Ukraine. Kharkiv. KhNADU.

Vasetsky Yu.M., Dziuba K.K. 2017. Analytical calculation method of quasi-stationary three-dimentional electromagnetic field created by the arbitrary current contour that located near conducting body. *Tekhnichna Electrodynamika. № 5*: 7-17.

Wohlfahrt H. 2009. New calculations checking an adequate materials law. New results on distortion measurements. Report on the Round Robin Tests on Residual Stresses. *International Institute of Welding Document* No IIW-X-1668-09, IIW-XIII-2291-09, IIW-XV-1326-09.

Wohlfahrt, H., Nitschkepagel, T., Dilger, K. et al. 2012. Residual Stress Calculations And Measurements — Review And Assessment Of The Iiw Round Robin Results. *Weld World*. **56**: 120–140.

Welding and Related Technologies – Krivtsun et al. (Eds)

Fatigue behavior of thin sheet joints of MA2-1M magnesium alloy produced by TIG and FSW

S.I. Motrunich, A.G. Poklaytskiy, I.M. Klochkov, O.A. Katok, M.O. Tsysar, P.M. Tkach, V.M. Bulash & V.V. Knysh
E.O. Paton Electric Welding Institute of the National Academy of Sciences of Ukraine, Kyiv, Ukraine

ABSTRACT: This study investigates the fatigue behavior and mechanical properties of welded joints of magnesium alloy MA2-1M produced by Friction Stir Welding (FSW) and Tungsten Inert Gas (TIG) welding. The study concludes that while TIG-welded joints with reinforcement demonstrated the highest ultimate tensile strength, approximately 96% of the base material's strength, compared to FSW joints. But FSW joints provides more uniform and reliable welds with superior fatigue performance with only a 7% reduction compared to the base metal at 10^6 cycles. These findings contribute to the optimization of welding techniques for critical magnesium alloy structures, particularly in high-performance applications such as aerospace and automotive industries.

1 INTRODUCTION

Magnesium alloys belong to the class of light metals and are highly valued for their low density and excellent strength-to-weight ratio, making them ideal for weight-sensitive applications. These alloys offer ultimate tensile strengths of up to 290 MPa and can maintain their structural integrity at temperatures as high as 450°C (Huisheng 2024). As a result, they are extensively employed in the aerospace and automotive industries, where weight reduction is critical for improving fuel efficiency and performance.

Magnesium alloys can be alloyed with elements such as aluminum, zinc, and rare earth metals to further enhance properties like corrosion resistance, ductility, and creep strength. Their lightweight nature also makes them a valuable material for portable electronics, sports equipment, and even biomedical devices.

However, one of the challenges of working with aluminum and magnesium alloys is their susceptibility to cracking and oxidation during welding processes. To address this, specialized fusion welding methods such as tungsten inert gas (TIG) welding, laser welding, and friction stir welding are often used to produce non-separable joints while minimizing defects (Vyskoč 2024, Poklaytsky 2014). These welding techniques help preserve the mechanical properties and improve the reliability of magnesium alloy components in high-stress environments. During fusion phase welded joints formed as a result of melting the edges of the connecting parts and filler wire in a common welding pool with their subsequent crystallization. Resulting in cast macrocrystalline structure, which leads to a decrease welded joints mechanical property compared to the base material (Srinivasan 2024). At the same time, it is possible to avoid metal melting in the welding zone and preserve the properties of the main material in the welded joints as much as possible thanks to the use of welding in the solid phase by friction stir welding (FSW) (Pasetti-Roza 2024, Poklyatskyi 2024). As a result of heating due to friction in the welding zone of a certain volume of connected parts to a plastic state, its intensive mixing, deformation and compaction by the working surfaces of the tool, a finely dispersed structure is formed in the joint (Khokhlov 2024, Lukinenko 2020). Thanks to this, welded joints obtained by FSW have high mechanical properties (Prihajatno 2024). However, for assessing

DOI: 10.1201/9781003518518-37

the performance of welded structures, the fatigue behavior of welded joints during their operation is of great importance.

The main purpose of the work is to study mechanical properties and fatigue behavior of welded joints of MA2-1M magnesium alloy, obtained by TIG and FSW.

2 WELDING PROCEDURE

Thin sheets of structural magnesium alloy MA2-1M of the Mg-Al-Zn-Mn alloying system were used for study. In particular it has following amount of alloying elements: 4,5% Al, 0,95% Zn, 0,47% Mn, 03% Fe, 0,06% Si, 0,01% Cu. The mechanical properties of thin shits (2 mm thickness) are shown in the Table 1.

Table 1. Mechanical properties of sheets of structural magnesium alloy MA2-1M with a thickness of 2 mm.

Mechanical properties	Rolling Direction	Transvers Direction
Tensile strength (σ_m), MPa	274...279	279...283
Yield strength ($\sigma_{0,2}$), MPa	179...200	203...219
Elongation (δ_5), %	19,5...20,3	23,5...26,3
Bending angle (α), degree	32...38	38...45

TIG welding was carried out using the ASTV-2M installation from the MW-450 power source (Fronius, Austria). In order to create favorable conditions for intensive mixing of the molten metal of the welding pool and to reduce the probability of the formation of pores in the seams, a current with a rectangular wave shape with a frequency of 200 Hz was used. The diameter of the tungsten electrode was 3.2 mm, and its length was 4 mm. The welding speed was 20 m/h, In order to obtain a chemical composition of the base metal in the weld, a band of base material with a thickness of 2 mm was used as a filler material to form the reinforcement of the joint. It was fixed between the welded edges by lowering grooves 1 mm deep and 4 mm wide to the bottom. The height of filler metal was 6...7 mm, which ensured the formation of the necessary weald reinforcement. Directly before welding, the welded edges (from three sides) and bands of filler material (from four sides) were mechanically cleaned to a depth of 0.10...0.12 mm. To ensure reliable protection of the metal from oxidation in the welding zone, high-grade argon shielding gas was used. At the same time, the inner diameter of the protective ceramic nozzle was 16 mm, and the gas consumption was 20...22 l/min. High-quality weld formation under such welding conditions was ensured at a current of 145...150 A.

Friction stir welding was performed on a laboratory equipment developed by E.O. Paton Electric Welding Institute. It provides a constant tool rotation frequency of 1420 rpm. Butt joints were obtained at welding speeds of 20 m/h., to reduce the probability of the formation of internal defects in the seams in the form of cavities inherent in this process (Ishchenko 2010). For welding, we used a special tool of our own design with a blade diameter of 12 mm and a tip length of 1.85 mm in the form of a truncated cone with a diameter of 3.4 mm at the base of the blade. To prevent possible surface contaminants and oxide films from entering the seam formation zone, mechanical cleaning of the welded edges (from three sides) to a depth of 0.10...0.12 mm was carried out immediately before welding.

3 RESULTS OF EXPERIMENTAL STUDIES

All mechanical testing was performed on servohydraulic testing systems MTS 318.25. Tensile testing was performed according to ISO 4136. Specimens (of 15 mm test portion width) to determine the tensile strength of welded joints were produced. The bending testing was performed according to ISO 5173. The bending angle was determined on samples of welded joints with a width of 10 mm. Loading was applied at the weld root of the joint. Fatigue testing was performed according ISO 1099 using force-controlled method. The width of the specimens for fatigue test was 25 mm. The fatigue tests were carried out either to fracture with stress ratio R=0.1 and 15 Hz frequency. A batch of 5...6 specimens of the same type were tested under the same conditions. The experimental data of fatigue tests were processed by a linear regression analysis, generally accepted for this type of testing. The corresponding S-N curve was built based on the results of tests for each batch of specimens and guidance of standards and recommendations.

As a result of tensile testing, it was established that the FSW joints fractured in the thermo-mechanical affected zone (TMAZ) from the retreating side (Figure 1a). Fracture occurs at weld face at a distance of 2.5...3.0 mm from the border of thermal influence, and from the weld root - in the zone of fusion of the weld with the base material. The ultimate strength of the FSW samples of butt joints obtained is at the level of 234...236 MPa.

As welded TIG samples with weld reinforcement, have ultimate strength up to 267 MPa, which is 96% of this indicator of the base material. Fracture of as welded TIG joints with reinforcement occurs from the weld face along the zone of fusion with the base material (Figure 1b), and from the weld root - along the main metal in the zone of thermal influence at a distance of about 5 mm from the fusion zone(Figure 1d). TIG samples with removed reinforcement to the level of the main material fracture in weld metal and shows a ultimate strength up to 233 MPa. At the same time, fracture occurs at the weld face near central part (Figure 1c), and from the weld root - in the fusion zone with the base material (Figure 1e).

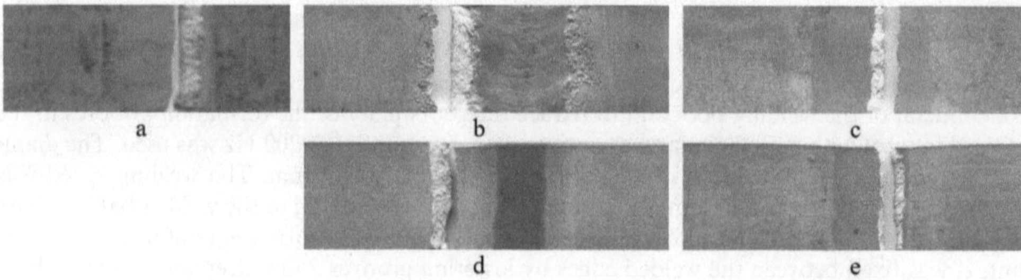

Figure 1. Samples fracture after tensile testing: a – FSW joints, b, c – as welded TIG joints with reinforcement, c,e – machined TIG joints without reinforcement, a, b, c – weld face, d, e – weld root.

Experimentally obtained S-N curves of butt-welded joints of magnesium alloy MA2-1M made by FSW demonstrate a high level of fatigue strength. Fatigue strength of FSW joints only 7% lower at 106 cycles and 20% at 105 cycles of the respective values of the base metal (Figure 2). Fatigue strength in high cyclic region $5 \cdot 10^4$... 10^6 cycles for machined TIG welds without reinforcement are 16...26% higher than corresponding values of as welded TIG joints. Fatigue strength of as welded FSW joints at maximum alternate stress at 10^6 cycles reach 106 MPa and TIG joints without reinforcement reach up to 89 MPa.

The conducted bending experimental studies showed that the angle of the samples of welded joints obtained by TIG is 76° on average and varies within 74°...78°. A crack initiation in the process of testing occurs in the central part of the seam. The maximum bending angle of samples obtained by FSW reach 64°...70°. At the same time, the crack is located on the weld face in the zone of thermomechanical influence near (2.5...3.0 mm) the zone of thermal influence, and from the weld root - in the zone of conjugation of the weld with the base material, where the maximum strengthening of the metal occurs during the welding process.

4 CONCLUSION

1. FSW and TIG welding demonstrated different mechanical behaviors in welded magnesium alloy joints. FSW joints exhibited superior uniformity and reduced defect formation in the

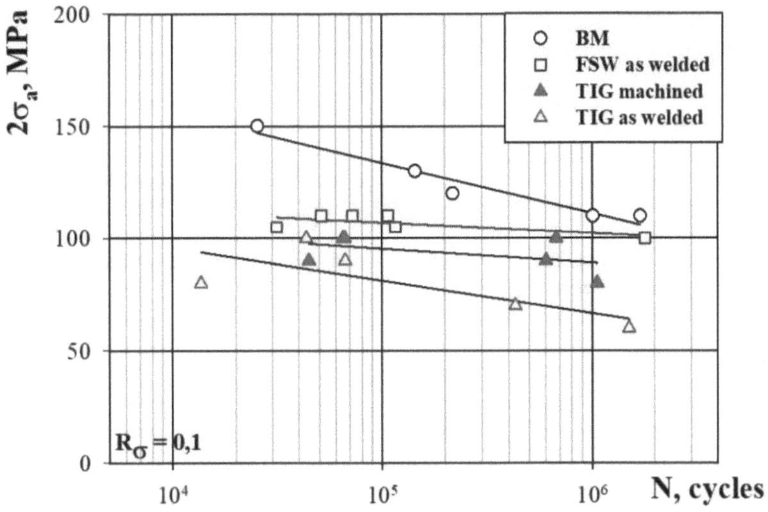

Figure 2. S-N curves of base metal and welded joints of MA2-1M magnesium alloys at stress ratio R = 0,1: o – base metal; □– as welded FSW joint; ▲ – machined TIG joint without reinforcement, Δ – as welded TIG joint with reinforcement.

thermomechanical affected zone (TMAZ), resulting in more reliable welds under fatigue testing. However, TIG welding produced higher ultimate tensile strength, particularly in as-welded samples with reinforcement, but was more prone to defect formation that affected fatigue performance.

2. Tensile testing shows that TIG-welded joints with reinforcement has highest ultimate strength at 267 MPa, equivalent to 96% of the base material, while FSW joints achieved 234–236 MPa. Bending tests revealed that TIG joints could achieve bending angles of 74°–78°, which exceeded those of FSW joints (64°–70°). Fracture in both cases typically occurred in the fusion zone or TMAZ, highlighting the critical importance of weld zone integrity.
3. Fatigue testing showed that FSW joints have a high fatigue strength, with only a 7% reduction compared to the base metal at 10^6 cycles, while TIG joints, particularly those without reinforcement, demonstrated lower fatigue resistance. TIG joints without reinforcement exhibited fatigue strengths 16%–26% higher than their as-welded counterparts but still lagged behind the FSW joints, which maintained a maximum alternating stress of 106 MPa at 10^6 cycles, compared to 89 MPa for TIG welds without reinforcement.

REFERENCES

Ishchenko, A. and Poklyatsky, A. 2010. Tool for friction stir welding of aluminium allous. Pat. 54096 Ukraine, Int. Cl. B23K 20/12; u201005315; fill. 30.04.2010; publ. 25.10.2010 [in Ukrainian].

Huisheng Cai et al. 2024. Effects of cooling rate on the microstructure and properties of magnesium alloy-a review, *Journal of Magnesium and Alloys*. https://doi.org/10.1016/j.jma.2024.06.021.

Khokhlov, M. et al. 2024. Thermo-mechanical processes in friction stir welding of magnesium alloy sheets, *The Paton Welding Journal*, 5:11–17.

Lukinenko, A. et al. 2020. Noise Level Assessment and Mechanical Properties of Welded Joints of Aluminium Alloys of the Al- Cu-Li System in FSW and TIG Welding, *FME Transactions*, 49(1):220–224.

Pasetti-Roza, A. et al. 2024. Behavior of microstructure and mechanical properties in the stir zone of friction stir welded ME21 magnesium alloy, *Journal of Materials Research and Technology*,29:4895–4901.

Poklaytsky, A. et al. 2014. Structure and properties of AMg2M alloy joints made by argon nonconsumable–arc welding and friction stir welding. *Applied Mechanics and Materials*. 682:166–169.

Poklyatskyi, A. et al. 2024. Strength and structure of MA2-1M magnesium alloy butt joints produced by argon arc welding with a non-consumable electrode and by friction with stirring., *The Paton Welding Journal*, 2:14–18.
Prihajatno, M. et al. 2024. Tensile and fatigue properties of friction stir AZ31B-H24 magnesium alloy welded joints under the influences of pin geometry and tool rotation rate, *Journal of Advanced Joining Processes*, 9:100202,
Srinivasan, S. et al. 2024. Effect of pulse gas tungsten arc welding on the metallurgical characteristics & microstructural evolution in AZ31B magnesium alloy welds. *Journal of Materials Research and Technology*, 29:5412–5424.
Vyskoč, M. et al. 2024. Microstructure – Properties relationship in laser-welded AZ31B magnesium alloy, *Materials Characterization*, 208:113664.

Welding and Related Technologies – Krivtsun et al. (Eds)

Corrosion fatigue of duplex stainless steel 2205 welded with magnetic field

O. Bilyy, R.B. Chi Bojórquez & G. Domínguez Rodríguez
Centre for Corrosion Research Autonomous University of Campeche

C.A. León Gomez
University of Sciences and Arts of Chiapas

ABSTRACT: Duplex stainless steel 2205 is an alloy containing 22% Cr and 5% Ni. The steel microstructures are a δ-ferrite matrix with BCC structure with intercalated FCC lattice with γ-austenite in the ratio 50:50. Duplex stainless steel is used in industry for its high mechanical and corrosion resistance, better than in ferritic or austenitic stainless steels. During the welding process a low intensity electromagnetic interaction was induced using axial external electromagnetic fields 0, 3 and 12 mT. Resistance to crack formation and growth was also evaluated in a low cycle corrosion fatigue test mode to monitor short crack behavior. It was found that low intensity electromagnetic interaction (3 mT) improves localized corrosion resistance.

1 INTRODUCTION

Duplex Stainless Steel 2205 is an alloy with a chemical composition with Cr (22%), Ni (5%) and Mo (3%) as main alloying agents. The mixed microstructure of Duplex Stainless Steel consists of austenite (γ) phase islands with a face-centered cubic crystal structure on a ferrite (δ) phase matrix, which has a body-centered cubic structure in a percentage ratio of approximately 50/50. Duplex Stainless Steels are widely used in the paper, chemical, petrochemical, energy and construction industries, to mention a few cases, since it has been observed that they have greater mechanical strength and corrosion resistance than austenitic or ferritic stainless steels separately (Strubbia et al. 2012), highlighting their superior resistance to corrosion and stress cracking when used in solutions containing chlorides and under conditions with acidification of the medium due to stagnation (Nilsson 1992). Crack behavior in engineering components and structures built with Duplex Stainless Steel is strongly influenced by the mode and type of loading, and microstructural features including alloy chemical composition, phase volume fraction, distribution, grain size, and heat treatment (Miller 1987). Although some researchers have proposed low intensity electromagnetic interaction during arc welding to minimize the precipitation of undesired phases, as well as the redistribution and change of the δ/γ phase ratio, which greatly improved localized corrosion resistance as it suppressed the precipitation of σ and χ phases, as well as the formation of chromium carbides and nitrides (García-Rentería et al. 2014). This research work evaluates the effect of applying an external magnetic field of 3 mT during arc welding process of duplex stainless steel 2205 on the microstructural evolution in increasing the resistance to fatigue corrosion cracking.

DOI: 10.1201/9781003518518-38

2 METHODOLOGY

2205 DSS plates (6.35 × 70 × 150 mm) were welded with a single V-groove preparation, using the arc welding process with a 1.2 mm diameter ER-2209 electrode (160 mm/s) and a shielding gas mixture of 98% Ar + 2% O2 (17 L/min). The axial magnetic fields applied during welding were 0 and 3 mT. The welding torch was traveled at 3.6 mm/s with a 10 mm rod. The current (236 to 28 A) and voltage (27.5 V) were adjusted to maintain an approximate heat input of 1, kJ/mm considering an efficiency of 75% for the arc welding process. For sample preparation, cross sections were made in the direction of the weld bead and machined into a prism with a length of 5 mm, a height of the same measurement and a length of 10 cm, containing the weld bead in the center, as shown in Figure 1. The rectangular shape of the specimen with the geometric orientation was used for the corrosion fatigue test with dimensions of 5 mm x 5 mm x 100 mm. A sinusoidal loading waveform with a frequency of 0.90 Hz was used. The initial maximum stress of 0.99 σys (yield limit) and a load ratio of 0.5 was applied for the fatigue test. The loading was parallel to the short transverse direction. The fatigue tests were performed in natural seawater at open circuit potential, all at room temperature. After determining the fatigue life, some interrupted fatigue tests were also performed at certain loading cycles.

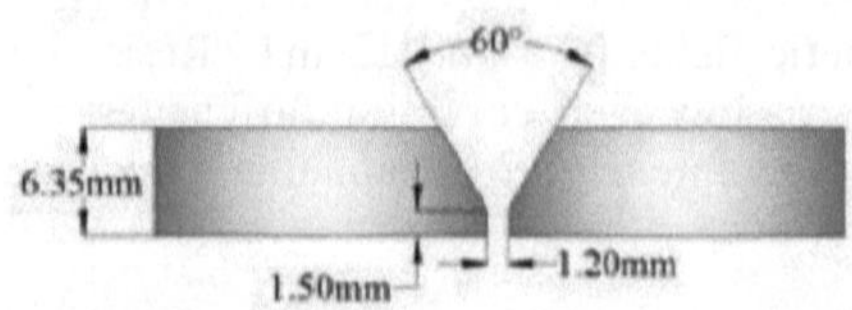

Figure 1. Simple V preparation of welded joints.

3 RESULTS

Four cases of possible cracks are considered as potential defects. Such analytical basis serves to assess the durability of roads with defects such as cracks by the factor "resistance of the building element to crack growth" which is the characteristic of the rate of change SIF K_I near the tip of the crack of a length during its growth in a specific structural element. Critical values of crack depth were determined: 1) when the crack depth is less than its safe value, at which no further crack propagation is observed $a<a_{th}$, where a_{th} - threshold size of the crack; 2) when the crack depth is greater than the value at which spontaneous crack growth is possible, that leads to brittle failure of the research object: $a>a_{fc}$, where a_{fc} - critical size of the crack (Panasyuk et al. 2014). To implement this approach, the extent of the investigated plate with such existing defects as existing cracks is considered: extension of the plate with centered cracks through the thickness; extension of the plate with extended surface crack; extension of the plate with two extended surface cracks; extension of the plate with semi-elliptical surface crack; extension of the plate with quarter-elliptical corner crack; extension of the plate with embedded (centered) elliptical crack, the models of which are shown in Figure 2a–d.

It should be noted that in cases, 3 and 4, where the crack is established by various ellipse-shaped figures, changing the ratio of the crack ellipse axes as a defect of different geometry.

In the investigation, the corresponding dimensions less dependencies for the considered cases shape and location based on the given and constructed analytical relations:

$$\frac{\sqrt{t}}{\sigma}\frac{dK_I}{da} = F\left(\frac{a}{t}\right), \tag{1}$$

where t – size of the structural element in the direction of crack propagation; σ– applied load, a – crack depth, which is a variable value compared to t which is constant.

Therefore, in a further study of the results of the research, we will introduce the variable parameter $(a/t)_*$ that characterizes the effective size of the defect. It should be noted that the characteristic feature of these dependencies is that a certain value of the parameter $(a/t)_*$, is always observed in them, starting from this value, a sharp increase in the stress intensity coefficient occurs.

This value $(a/t)_*$ was considered characteristic for evaluating the strength and reliability of structural elements with defects such as cracks.

With the crack growth data, the statistical analysis of Paris dependencies was carried out for each of the cases. In Figures 3-6 we can see that 0 mT and 3 mT have a similar behavior, 3 mT being much better, on the other hand, welding with a magnetic field of 12 mT presents an unfavorable behavior even in worse conditions than conventional welding. In sea water, 0 mT and 3 mT welds have a very similar behavior until the stress is reduced to 90% of the yield limit, where the effect of the electrolyte is more evident and the electrochemical characteristics previously shown by 3 mT welds make a difference in favor of it, being more resistant to crack initiation.

For the determination of the characteristic values of the crack length in the studies of structural elements, the experimental basis and the analytical relationship were used: $dc/dN = C(\Delta K)^n$, where C and n - constant characterization system “material - environment” (Table 1).

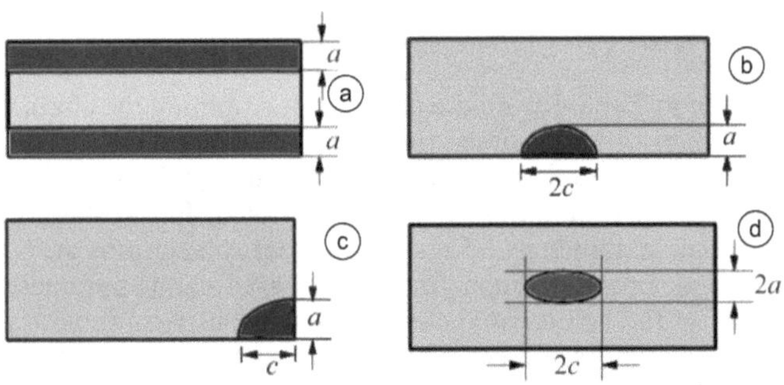

Figure 2. Models of the investigated plate with such existing defects as cracks: (a) plate extension with two extended surface cracks; (b) plate extension with semi-elliptical surface crack; (c) plate extension with quarter elliptical corner crack; (d) plate extension with embedded (centered) elliptical crack.

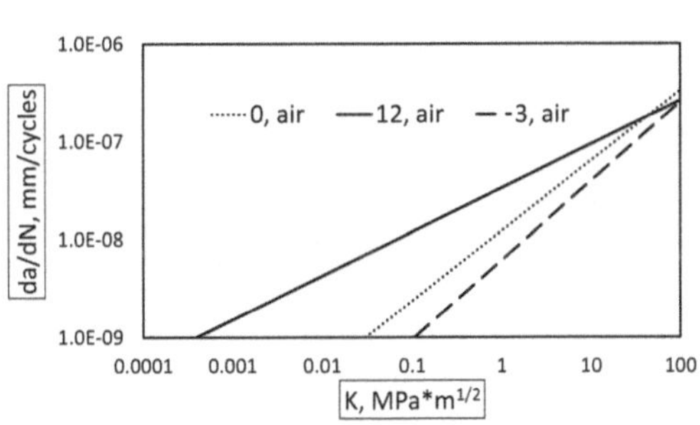

Figure 3. Paris dependence for different materials in air: 12 mT; 0 mT; 3 mT.

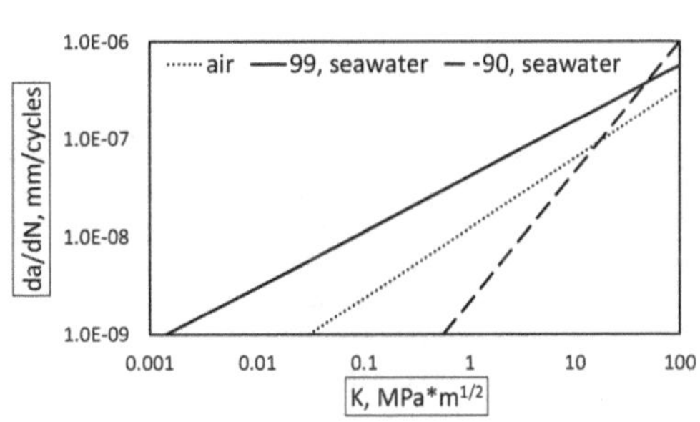

Figure 4. Paris dependence for 0 mT in different environments and conditions: 99% of the yield limit, seawater; 99% of the yield limit, air; 90% of the yield limit, seawater.

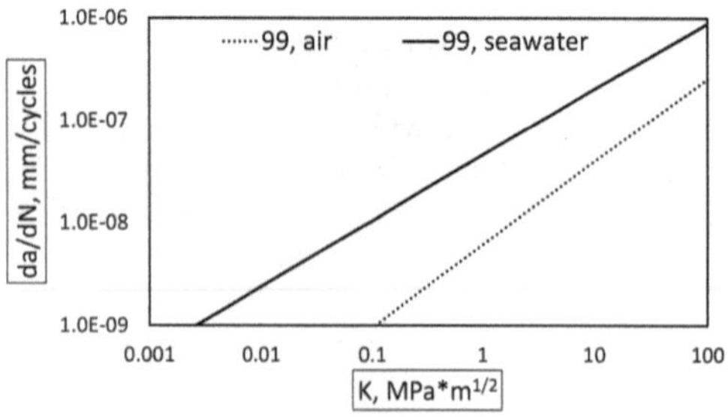

Figure 5. Paris dependence for 3 mT in different environments and conditions: 99% of the yield limit, air; 99% of the yield limit, seawater.

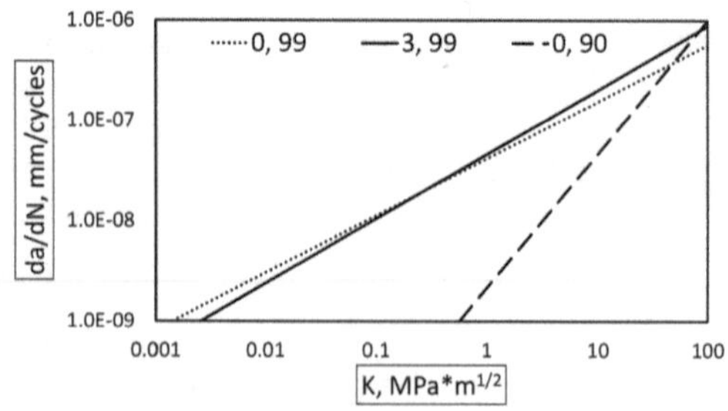

Figure 6. Paris dependence in seawater: 99% of the yield limit, 3 mT;; 99% of the yield limit, 0 mT; 90% of the yield limit, 0 mT.

Table 1. "Material-Environment" constants and critical SIF values.

	Constants of the "material – environment" system		
"Material – environment" system	C	n	ΔK_{th}, $MPa\sqrt{m}$
0 mT – air	1.9E–8	0.582	17.285
0 mT – seawater	4.16E–8	0.568	4.675
3 mT – seawater	3.18E–8	0.677	5.448

The ΔK_{th} values given in Table 1 were the basis for determining the characteristic values a_{th}. Together with these values obtained based on processing experimental data, the values of a_{fc} were calculated using the technique of (Panasyuk et al. 2014), which corresponds to the value of the characteristics $(a/t)_*$. In addition, it should be noted that in the investigated object 5 x 5 x 5 mm was the angle parameter of 5 mm, all further calculations are given for this case, when the crack propagates perpendicular to the main axis of this parameter. In this work it evaluate the durability of the investigated element of the construction with possible cracks as possible defects of various shapes and geometry, which provides the conditions of achievement of spontaneous destruction of structural elements, i.e.

$$N_{fc} = \int_{a_{th}}^{a_{fc}} \frac{da}{F(\Delta K_I)}, \tag{2}$$

where N_{th} - number of loading cycles before destruction of structural elements; $F(\Delta K_I)$ - function of SIF. The results of the durability calculation are given in Table 2

Table 2. Data for calculating the durability of the research object.

Model	"Material – environment" system	a_{th}, mm	a_{fc}, mm	$N_{fc}, cycles$
Model 1	0 mT – air	0.20	2.3187	1.27968E+7
	0 mT – seawater	0.015		7.31996E+6
	3 mT – seawater	0.02		6.8981E+6
Model 2	0 mT - air	0.22	1.67381	9.1577E+6
	0 mT – seawater	0.01		5.93539E+6
	3 mT – seawater	0.02		5.33695E+6
Model 3	0 mT – air	0.20	2.18112	1.166936E+7
	0 mT – seawater	0.01		6.790957E+6
	3 mT - seawater	0.02		5.33695E+6
Model 4	0 mT - air	0.20	1.78798	9.76859E+6
	0 mT – seawater	0.015		6.2387E+6
	3 mT - seawater	0.02		5.6237E+6

From the results, presented in Table 2, the corrosion environment strongly influences the durability of the investigated structural element, and such a trend is observed for all considered cases, as well as for all modelled defects such as cracks. In general, the durability of the investigated object in air is higher. As for the environment, the durability is somewhat better in the case of 0 mT, although with a very insignificant difference.

3 CONCLUSIONS

The crack propagation rate in the 3 mT weld is lower than that in the 0 mT weld, indicating that the fracture toughness of the 3 mT weld is higher compared to conventional welding. Although the difference in the time elapsed for crack nucleation and the propagation rate is notable when comparing the conditions of the welded joints, differences were observed in the mechanism of fatigue crack nucleation and propagation in the specimens of both conditions. Based on experimental research, the value of SIF criterion for duplex stainless steel 2205 is determined. By applying analytical relations for SIF characteristics, the value of crack is calculated as length defects that may occur in the considered frame construction. Engineering calculations are suggested for the evaluation of the durability of the given object and are presented with modeled cracks as defects of various shapes and geometries based on the number of loading cycles.

REFERENCES

García-Rentería M. A., López-Morelos V.H., García-Hernández R., Dzibpérez L., García-Ochoa E.M., González-Sánchez J. Improvement of Localised Corrosion Resistance of AISI 2205 Duplex stainless Steel Joints Made by Gas Metal Arc Welding Under Electromagnetic Interaction of Low Intensity.// Applied Surface Science 321, 2014, P. 252–260.

Miller KJ. The Behaviour of Short Fatigue Cracks and their Initiation part I- A review of two recent books. Fatigue & Fracture of Engineering Materials & Structures, 10, 1987, P. 75–91.

Nilsson J.O.. Overview Super duplex stainless steels. // Materials Science and Technology, 8, 1992, P. 685–700.

Panasyuk V.V., Dmytrakh I.M., Toth L., Bilyi O.L., and Syrotyuk A.M. A method for the Assessment of the Serviceability and Fracture Hazard for Structural Elements with Crack-like Defects/// Materials Science (Springer Science + Business Media New York). – 2014. – V. 49, No 5. – P. 565–576.

Strubbia R., Hereñú S., Marinelli M.C., Alvarez-Armas I.. Short crack nucleation and growth in lean duplex stainless steels fatigued at room temperature. International // Journal of Fatigue, 41, 2012, P. 90–94.

Welding and Related Technologies – Krivtsun et al. (Eds)

Comprehensive analysis of the microstructure and mechanical properties of deposited metal of nickel high-temperature alloy ZhS32

Hanna Zviahintseva, Oleksandr Yarovytsyn, Mykola Cherviakov & Iryna Sviridova
Paton Welding Institute of NAS of Ukraine, Kyiv, Ukraine

ABSTRACT: During industrial use of the microplasma powder metal deposition (MPD) process at the E.O. Paton Electric Welding Institute, positive experiences with Ni-based superalloy ZhS32 on aircraft blades were confirmed. A comprehensive analysis, including optical and scanning electron microscopy, and mechanical property studies, aimed to improve the repair area's characteristics and expand the range of parts repairable by MPD. The study examined the ZhS32 alloy's structural states after deposition, aging and homogenization heat treatments.

1 INTRODUCTION

For two decades, the domestic aircraft repair industry has employed microplasma powder welding deposition (MPWD) to restore the shroud edges of D-18T gas turbine engine blades (Yushchenko et al. 2010, Zhemanyuk et al. 2015, Yushchenko et al. 2022). During this period, the nickel-based superalloy ZhS32 (Table 1), used in the "base-deposited metal" joint, has shown high sensitivity to heat input, leading to crack formation (Yushchenko et al. 2010, Yushchenko et al. 2022). This alloy, which is challenging to weld according to the generally accepted classification (Sims et al. 1987, DuPont et al. 2009) due to its 62% γ'-phase content (Kuznetsov et al. 2010), requires further investigation.

This study aims to determine the causes of hot crack formation in ZhS32 welded joints by comparing the short-term mechanical properties and microstructure of the base and deposited metal.

2 EXPERIMENTAL SETUP

MPWD was performed on "vertical wall" workpieces of ZhS32 superalloy with these parameters: pulsed current I_{RMS} = 20-35 A, voltage U = 22-24 V, powder feed rate G_p = 2.5-5.0 g/min, and torch speed v = 0.8-1.5 m/hr. The substrate was a 3.0 mm wide AISI 304 stainless steel plate edge. Powder of ZhS32 superalloy with 63-160 μm fraction spherical-shaped particles was used as additive material. Argon was used as the plasma and transporting gas (0,5-1,5 l/min and 3-6 l/min respectively), while a 90% Ar + 10% H_2 mixture served as the shielding gas (5-10 l/min).

Each deposited layer was air-cooled to room temperature before mechanical processing with abrasive tools to remove oxidized surfaces. After grinding the deposited metal to a thickness of 2.5-3.5 mm, liquid penetrant tests (LPT) confirmed the absence of defects.

Table 1. Chemical composition of ZhS32 nickel-based superalloy.

	wt, %										
Alloy	C	Cr	Al	Ni	Mo	W	Co	Re	Ta	Nb	V	Fe
ZhS32	0.12 0.18	4.3 5.6	5.6 6.3	Bal	0.8 1.4	7.8 9.5	8.0 10.0	3.5 4.5	3.5 4.5	1.4 1.8	≤0,15	≤0,15

DOI: 10.1201/9781003518518-39

The samples were cut via wire EDM, and subsequent LPTs verified defect-free surfaces after final machining and heat treatment. The choice of shape and size of miniature flat proportional samples is preliminarily justified in work (Yushchenko et al. 2019). The tensile tests, conducted on an MTS-810 machine at 20 - 1100°C, evaluated short-term strength parameters, including ultimate tensile strength (UTS) or tensile strength Rm and critical fracture strain (ε_{cr}) or percentage total extension at fracture A_t.

Mechanical properties were assessed in the following conditions: "as built" (after MPWD without heat treatment), after heat treatment (HT) at 1050°C-2.5 h according to the aging mode (Yushchenko et al. 2010), and after homogenization (Kuznetsov et al. 2010) at 1280°C-2.5 h. Base metal samples were similarly tested after homogenization. The specified plates (50×100×5 mm) were produced at the SE "Ivchenko - Progress" on the industrial technology of vacuum-induction melting with a directional and polycrystalline structure. Metallographic studies using optical and scanning electron microscopy (SEM) in back-scattered electron (BSE) were conducted to analyze microstructure formation in the ZhS32 base and deposited metals.

3 EXPERIMENTAL RESULTS AND DISCUSSION

Previous metallographic studies have revealed significant structural differences between the deposited metal (Figure 1a) and the cast base metal (Figure 1b) of ZhS32 superalloy, particularly in the dendrite spacing. However, metallographic analysis of deposited metal samples, which were fractured during high-temperature longitudinal tensile tests, demonstrated an intergranular fracture mode. Experimental tensile test data for ZhS32 deposited metal in various structural states, as well as for the cast base metal with directional and polycrystalline structures, are presented in Figure 2. SEM results for ZhS32 deposited metal are shown in Figures 3-4 and Table 2.

It was found that the deposited metal of ZhS32 superalloy exhibits reduced deformation capacity during tensile tests. In the "as-built" and post-aging states (1050°C - 2.5 h), the critical fracture strain (ε_{cr}) at 20-1100°C does not exceed 0.8%. Homogenization heat treatment improves plasticity at T≤600°C, but at T≥700°C, ε_{cr} remains between 0.35-1.5%. At T=900-1100°C, the homogenized deposited metal's ε_{cr} values approach those of the homogenized cast base metal with a polycrystalline structure, while the cast base metal with a directional

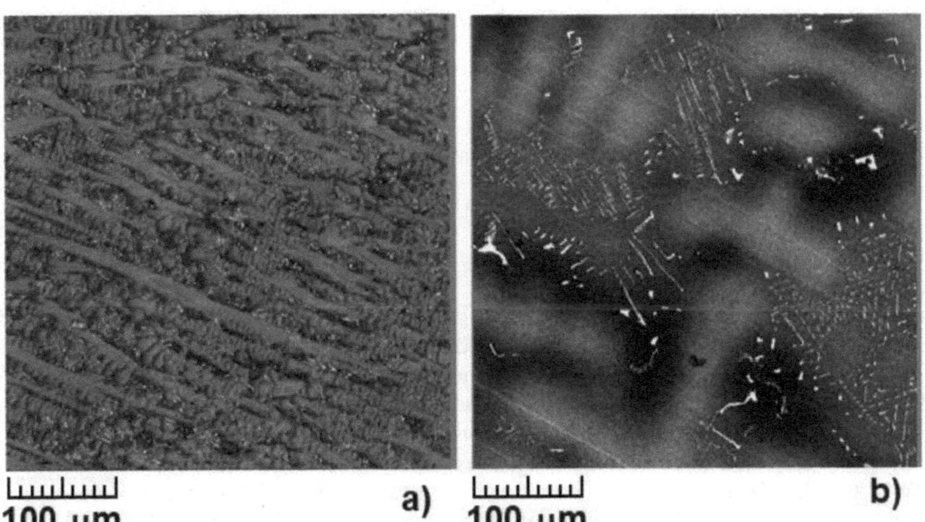

Figure 1. Features of the microstructure of the deposited (a) and cast (b) base metal of nickel-based superalloy ZhS32, SEM BSE.

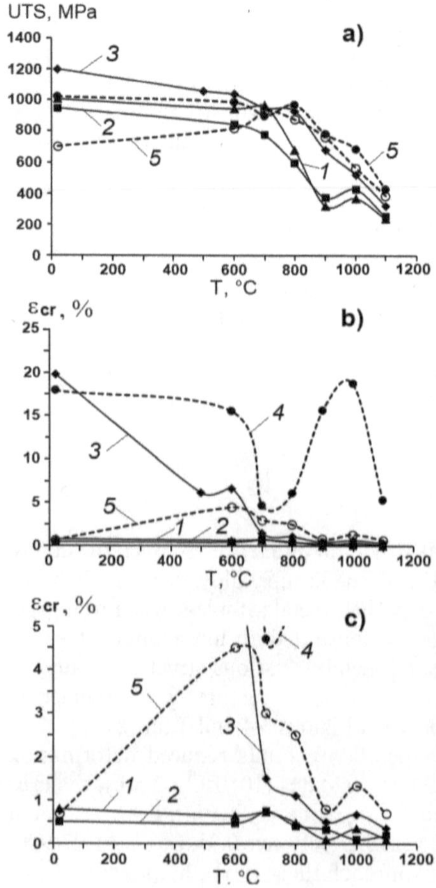

Figure 2. The complex of short-term mechanical properties of the deposited metal (DM) and base metal (BM) of ZhS32 superalloy: 1 – DM, "as built"; 2 – DM, after aging HT; 3 – DM, after homogenization HT; 4 – BM with a directed structure, after homogenization HT; 5 – BM with a polycrystalline structure, after homogenization HT. Notes: UTS – ultimate tensile strength; εcr – the critical fracture strain; c – given on a larger scale along the ε_{cr} axis.

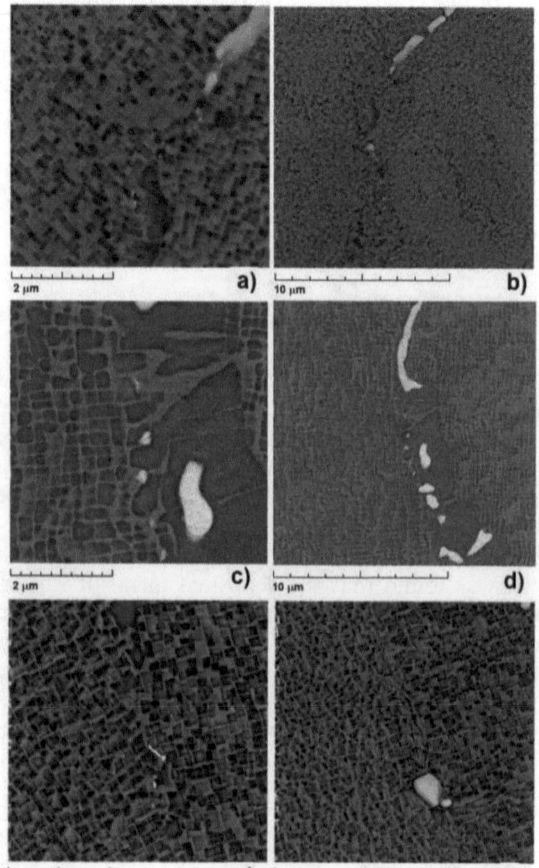

Figure 3. Precipitation hardening and grain boundary peculiarities of ZhS32 deposited metal, SEM BSE: a, b – "as built"; c, d – after aging HT; e, f – after homogenization HT.

structure shows the highest plasticity, with ε_{cr} ranging from 6.1-18.7% at T=20-1000°C with a minimum values at T=800°C.

At T≥800°C, the "as-built" deposited metal shows a 30-60% reduction in UTS compared to the homogenized base metal. Aging heat treatment (1050°C-2.5 h) slightly increases UTS at 1000°C. For homogenized ZhS32 deposited metal, UTS values at T=600-900°C are comparable to those of the homogenized cast base metal with a directional structure, and at T=800-1100°C, they are similar to those of the polycrystalline structure.

The study identified the causes of hot crack sensitivity during blade edge restoration using MPWD in industrial repair, linking it to the low ductility state (LDS) of the deposited metal at 20-1100°C in "as-built" and post-aging states. Metallographic studies indicate that LDS in ZhS32 deposited metal is due to the formation of a polycrystalline structure and carbon-enriched grain boundaries, leading to the formation of carbide phases and intergranular γ′-phase (Table 2).

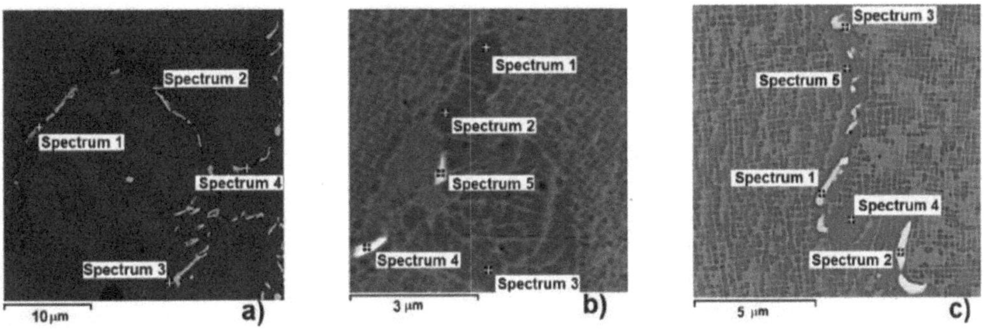

Figure 4. EDX-analysis of intergranular phases of ZhS32 deposited metal, SEM BSE: a, b – "as built"; c – after aging HT.

Table 2. Summary results on grain boundary EDX analysis of ZhS32 deposited metal (DM) in "as built" and in after aging HT structural states.

Specimen	Phase	Place of EDX-analysis	Average wt, %
DM ZhS32 "as build"	carbide	Figure 4a Spectrum 1-4	24,8 % C - 40,1% Ta - 24,4 % Nb - 7,0% W- 1,7 % Ni-1,0 % Mo - 0,6 % Cr
	carbide	Figure 4b Spectrum 4-5	25,8 %C - 34,5 % Nb - 21,5 % Ta - 8,9 % Ni -5,5% W - 1,4 % Mo - 1,0 % Cr - 0,7 % Co - 0,5 % Al
	γ '-phase	Figure 4b phase Spectrum 1	18,2 % C - 4,8 % Al - 4,3% Ta - 2,1 % Nb - 4,6 % W-57,1% Ni - 4,8 % Co - 2,0% Cr - 0,7 % Mo - 1,1 % Re
DM ZhS32 after aging HT	carbide	Figure 4c Spectrum 1	18,1 % C - 32,2% Ta - 20,7 % Nb - 10,8% W-11,5 % Ni -2,4% Co - 2,2 % Mo - 1,5 % Cr - 0,9 % Al
	carbide	Figure 4c Spectrum 2-3	11,6 %C - 26,6 % Re - 22,7 % W - 3,2 % Ta - 2,8 % Nb 3,7 % Mo -15,1% Ni - 9,2 % Co - 4,9 % Cr - 0,7 % Al
	γ '-phase	Figure 4c Spectrum 4-5	8,8 % C - 5,1 % Al - 4,3% Ta - 1,6 % Nb - 7,2 % W-61,3% Ni - 6,8 % Co - 2,3% Cr - 0,9 % Mo - 1,5 % Re

4 CONCLUSIONS

The short-term mechanical properties of ZhS32 superalloy's multi-layer deposited metal, produced via MPWD, were investigated at 20-1100°C using longitudinal tensile tests. Homogenization HT (1280°C - 1.5 h) enhanced the UTS of the deposited metal to 0.8-1.2 times that of the base metal. A key characteristic of ZhS32 deposited metal after homogenization HT during tensile tests at 20-1100°C is its limited deformation capacity, with ε_{cr} not exceeding 1.5% at 700-1100°C. The susceptibility of ZhS32 "base-deposited metal" welded joints to hot cracks is attributed to LDS with a ε_{cr} of no more than 0.8%, which forms during MPWD and persists after aging heat treatment (1050°C - 2.5 h). The LDS presence in the multilayer deposited metal ZhS32 is associated with the formation of carbon-enriched grain boundaries and the release along them of a significant amount of carbide phases with a dispersion of 0,2×0,2 to 0,5×5,0 μm, separated by gaps of the intergranular γ′-phase.

REFERENCES

DuPont, J.N., Lippold, J.C. & Kiser, S.D. 2009. *Welding Metallurgy and Weldability of Nickel-base Alloys*. John Wiley & Sons.

Kuznetsov, V.P., Lesnikov, V.P. & Konakova, I.P. 2010. *Structure and properties of high-temperature strength nickel alloy ZhS32-VI: Directory*. Yekaterinburg: Kvyst. (in Russian).

Sims, C.T., Stoloff, N.S. & Hagel, W.C. 1987. *Superalloys II. High-Temperature Materials for Aerospace and Industrial Power*. New York: John Wiley & Sons.

Yushchenko, K.A., Savchenko, V.S., Yarovitsyn, A.V., Nakonechny, A.A., Nastenko, G.F., Zamkovoj, V.E., Belozertsev, O.S. & Andrejchenko, N.V. 2010. Development of the technology for repair microplasma powder cladding of flange platform faces of aircraft engine high-pressure turbine blades. *The Paton Welding Journal* 8: 21–24.

Yushchenko, K.A., Yarovitsyn, A.V., Chervyakov, N.O., Zvyagintseva, A.V., Volosatov, I.R. & Khrushchov, G.D. 2019. Evaluation of short-term mechanical properties of a joint of difficult-to-weld nickel high-temperature alloys of ZhS6 type. *The Paton Welding Journal* 7: 29–35. DOI: http://dx.doi.org/10.15407/tpwj2019.07.07.

Yushchenko, K.A., Yarovytsyn, O.V., Khrushchov, G.D., Petrik, I.A. & Chygileichyk, S.L. 2022. Research and optimization of refurbishment of HPT blades of the D-18T aircraft gas turbine engine by micro-plasma powder welding. *Space Science and Technology* 3(28): 3–16. DOI: http://dx.doi.org/10.15407/knit2022.03. 01-01. (in Ukrainian).

Zhemanyuk, P.D., Petrik, I.A. & Chigilejchik, S.L. 2015. Experience of introduction of the technology of reconditioning microplasma powder surfacing at repair of high-pressure turbine blades in batch production. *The Paton Welding Journal* 8: 43–46. DOI: http://dx.doi.org/10.15407/tpwj2015.08.08.

Welding and Related Technologies – Krivtsun et al. (Eds)

Relationship between stress concentration and stress intensity factors in sharp or rounded V-notches in orthotropic plane

A.B. Chornenkyi, B.M. Datsko & S.M. Lavrys
Karpenko Physico-Mechanical Institute of the NAS of Ukraine, Lviv, Ukraine

ABSTRACT: By using the singular integral equations of the plane problem of the theory of elasticity for anisotropic body with cracks, integral equations for an orthotropic plane containing curvilinear holes with rounded vertices have been constructed. By using the limit transition when the radius of curvature of vertex tends to zero, the stress intensity factors in sharp vertices of the holes have been found.

1 INTRODUCTION

Karpenko Physico-Mechanical Institute of the NAS of Ukraine, known in Ukraine and in the world as a research center in the field of fracture mechanics and strength of materials (Khoma et al. 2022, Lavrys et al. 2023, Savruk&Kazberuk 2017). For instance, in fracture mechanics, considerable attention is paid to the investigation and study of anisotropic and orthotropic bodies with cracks. The models of anisotropic or orthotropic bodies are often used in the analysis of composite materials. Orthotropic materials are characterized by different properties in three mutually perpendicular directions.A characteristic feature of these materials is their high physical and mechanical properties in the specified directions.Currently, plane problems of elasticity theory for anisotropic bodies with rectilinear cracks are the most studied, their solutions can be obtained in the same way as for an isotropic medium.

2 BASIC RELATIONS OF THE PLANE PROBLEM OF ELASTICITY THEORY

The law of Hooke for a plane stress state for an orthotropic body can be represented in the form [1, 2]

$$\varepsilon_x = \frac{1}{E_x}\left(\sigma_x + \frac{\gamma_1^2+\gamma_2^2-2\gamma_3^2}{2}\sigma_y\right),\ \varepsilon_y = \frac{1}{E_x}\left(\gamma_1^2\gamma_2^2\sigma_y + \frac{\gamma_1^2+\gamma_2^2-2\gamma_3^2}{2}\sigma_x\right), \varepsilon_{xy} = \frac{\gamma_3^2}{E_x}\sigma_{xy},$$

where $\gamma_1, \gamma_2, \gamma_3$ – dimensionless parameters:

$$\gamma_{1,2} = \sqrt{\frac{E_x}{2G} - \nu_{xy} \pm \sqrt{\nu_{xy}^2 - \frac{E_x}{G}\nu_{xy} + \left(\frac{E_x}{2G}\right)^2 - \frac{E_x}{E_y}}},\ \gamma_3^2 = \frac{E_x}{2G}.$$

Here $E_x = E_1,\ E_y = E_2\ (E_x = E_2,\ E_y = E_1)$– elastic modulus for tensile-compression along the axes x and y, $\nu_{xy} = \nu_{12}\ (\nu_{xy} = \nu_{21} = \nu_{12}E_2/E_1)$ – Poisson's ratio of the compression of the plane in the direction of the axis $y\ (x)$ by stretching along the axis $y\ (x)$;

DOI: 10.1201/9781003518518-40

$G = G_{12} = G_{xy} = G_{21} = G_{yx}$ – a displacement module characterizing the change of angles between the principal axes.

For an elastic orthotropic body, the stress function satisfies the elliptic differential equation (Savruk&Kazberuk 2017, Kazberuk et al. 2016)

$$\frac{\partial^4 F}{\partial y^4} + (\gamma_1^2 + \gamma_2^2)\frac{\partial^4 F}{\partial x^2 \partial y^2} + \gamma_1^2\gamma_2^2\frac{\partial^4 F}{\partial x^4} = 0. \tag{1}$$

The corresponding characteristic equation

$$\mu^4 + (\gamma_1^2 + \gamma_2^2)\mu^2 + \gamma_1^2\gamma_2^2 = 0$$

has two pairs of complexly interconnected roots $\mu_k = i\gamma_k\ (\gamma_k > 0)$ and $\bar{\mu}_k = -i\gamma_k\ (k = 1, 2)$.

When the roots μ_1 and μ_2 are different, the total solution of the equation (1) can be expressed in the auxiliary complex planes through two analytic functions $\varphi_1(z_1)$ and $\varphi_2(z_2)$.

The singular integral equation for an orthotropic body (Savruk&Kazberuk 2017)

$$\frac{1}{\pi}\int_{L_1}[K_1(\tau_1, t_1)\phi_1'(\tau_1)d\tau_1 + L_1(\tau_1, t_1)\overline{\phi'_1(\tau_1)}d\bar{\tau}_1] = P_1(t_1),\ t_1 \in L_1, \tag{2}$$

Where kernels $K_1(\tau_1, t_1)$ and $L_1(\tau_1, t_1)$and the right-hand side $P_1(t_1)$ of the singular integral equation for an orthotropic body are determined by the relations

$$K_1(\tau_1, t_1) = \frac{i(\gamma_1-\gamma_2)}{2}\left(\frac{1}{\tau_1-t_1}\frac{dt_1}{dt} + \frac{1}{\bar{\tau}_2-\bar{t}_2}\frac{d\bar{t}_2}{dt}\right), L_1(\tau_1, t_1) = \frac{i(\gamma_1+\gamma_2)}{2}\left(\frac{1}{\bar{\tau}_1-\bar{t}_1}\frac{d\bar{t}_1}{dt} - \frac{1}{\bar{\tau}_2-\bar{t}_2}\frac{d\bar{t}_2}{dt}\right),$$
$$P_1(t_1) = \frac{1}{2}\left[(1+\gamma_2)p(t) - (1-\gamma_2)\overline{p(t)}\frac{d\bar{t}}{dt}\right].$$

3 RELATIONSHIP BETWEEN STRESS CONCENTRATION AND STRESS INTENSITY FACTORS IN ORTHOTROPIC PLANE

Let us consider an elastic orthotropic wedge with a vertex at the origin and a vertex angle $2\beta\,(0 \le \beta < \pi/2)$ (Figure 1).

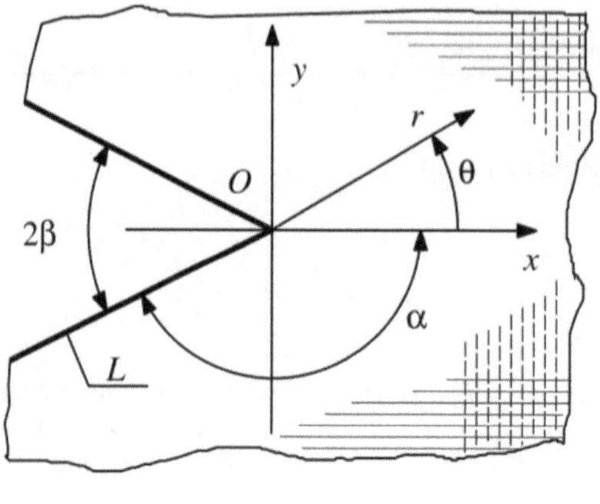

Figure 1. Orthotropic wedge with a vertex angle 2β.

We assume that the elastic wedge is under symmetric loading and occupies the area $S = \{(r, \theta);\ r \geq 0,\ -\alpha \leq \theta \leq \alpha\}$, where $\alpha = \pi - \beta$; r, θ – polar coordinates with the pole at the tip of the wedge and the polar axis along the diagonal of the wedge.

Cartesian coordinate axes are selected along the symmetry axes of the orthotropic elastic medium: $z = x + iy = re^{i\theta}$.

The distribution of singular stresses in a wedge with a load edge describes complex potentials (Savruk&Kazberuk 2017, Kazberuk et al. 2016)

$$\Phi_1^0(z_1) = \frac{R_2(\alpha)}{2[R_2(\alpha) - R_1(\alpha)]} \frac{\tilde{K}_{\mathrm{I}}^{\mathrm{V}}}{(2\pi z_1)^{\lambda_{\mathrm{I}}}},\quad \Phi_2^0(z_2) = -\frac{R_1(\alpha)}{2[R_2(\alpha) - R_1(\alpha)]} \frac{\tilde{K}_{\mathrm{II}}^{\mathrm{V}}}{(2\pi z_2)^{\lambda_{\mathrm{I}}}}.$$

Here $R_k(\theta) = (\cos^2\theta + \gamma_k^2 \sin^2\theta)^{1-\lambda/2} \cos(2-\lambda)\beta_k(\theta),\ k = 1, 2; \tilde{K}_{\mathrm{I}}^{\mathrm{V}} = \lim_{r \to 0} \left[(2\pi r)^{\lambda_{\mathrm{I}}} \sigma_\theta(r, 0)\right]$ – stress intensity factor in the tip of the wedge; λ_{I} – stress singularity exponents at the tip of a wedge, which is determined from the characteristic equation (Kazberuk et al. 2016)

$$(\gamma_2^2 - \gamma_1^2)\mathrm{tg}\alpha - \gamma_2(1 + \gamma_1^2 \mathrm{tg}^2\alpha)\mathrm{tg}[(2-\lambda)\beta_2(\alpha)] + \gamma_1(1 + \gamma_2^2 \mathrm{tg}^2\alpha)\mathrm{tg}[(2-\lambda)\beta_1(\alpha)] = 0, \tag{3}$$

where $\beta_k(\alpha) = \pi + \mathrm{arctg}(\gamma_k \mathrm{tg}\alpha)\,(\pi/2 < \alpha < \pi),\ k = 1, 2$.

Equation (3) in zone $0 < \mathrm{Re}\lambda < 1$ has no roots for the angles $0 < \alpha < \pi/2$, and for $\pi/2 < \alpha < \pi$ has one real root λ_{I} that belongs to the range $(0, 1/2)$. For $\alpha = \pi$ (half infinite crack) $\lambda_{\mathrm{I}} = 1/2$.

By making the limit transition when the parameter $\gamma_2 \to \gamma_1$, we obtain the characteristic equation for the quasi-orthotropic wedge ($\gamma_1 = \gamma_2 = \gamma$) (Ioakimidis&Theocaris 1977)

$$2\gamma \mathrm{tg}\alpha \cos[2(2-\lambda)\beta_1(\alpha)] - (1 - \gamma^2 \mathrm{tg}^2\alpha) \sin[2(2-\lambda)\beta_1(\alpha)] - 2\gamma(1-\lambda)\mathrm{tg}\alpha = 0.$$

The components of the stresses in the wedge are found by the formulas

$$\sigma_x^0 = -2\mathrm{Re}\left[\gamma_1^2 \Phi_1^0(z_1) + \gamma_2^2 \Phi_2^0(z_2)\right],\ \sigma_y^0 = 2\mathrm{Re}\left[\Phi_1^0(z_1) + \Phi_2^0(z_2)\right], \tau_{xy}^0 = 2\mathrm{Im}\left[\gamma_1 \Phi_1^0(z_1) + \gamma_2 \Phi_2^0(z_2)\right]. \tag{4}$$

By using the numerical solution of the integral equation (2) with kernels and the right part and the additional condition (4), we find the components of the stresses along the edges of the cut L by the formulas

$$\sigma_x^\pm = \sigma_x^0 - 2\mathrm{Re}\left[\gamma_1^2 \Phi_1^\pm(t_1) + \gamma_2^2 \Phi_2^\pm(t_2)\right],\ \sigma_y^\pm = \sigma_y^0 + 2\mathrm{Re}\left[\Phi_1^\pm(t_1) + \Phi_2^\pm(t_2)\right], \tau_{xy}^\pm = \tau_{xy}^0 + 2\mathrm{Im}\left[\gamma_1 \Phi_1^\pm(t_1) + \gamma_2 \Phi_2^\pm(t_2)\right],$$

where the stresses σ_x^0, σ_y^0, τ_{xy}^0 on the contour L are given by expressions (3), and the limit values of the potentials $\Phi_k(z_k)$ are given by the Sochocki-Plemel formulas

$$\Phi_k^\pm(t_k) = \pm\frac{1}{2}\Phi'_k(t_k) + \frac{1}{2\pi i} \int_{L_k} \frac{\Phi'_k(\tau_k)}{\tau_k - t_k} d\tau_k,\ t_k \in L_k,\ k = 1, 2.$$

Given that the edge of the notch is free of loads ($\sigma_n^* = \tau_{ns}^* = 0$) and the sum of normal stresses is invariant, that is $\sigma_n^* + \sigma_s^* = \sigma_x^* + \sigma_y^*$, we obtain(Savruk & Kazberuk 2017)

$$\sigma_s^*(\eta) = \sigma_x^0(\eta) + \sigma_y^0(\eta) + \sigma_x^-(\eta) + \sigma_y^-(\eta) = (2\pi\rho)^{-\lambda_{\mathrm{I}}} \tilde{K}_{\mathrm{I}}^{\mathrm{V}} R_{\mathrm{I}}(\eta),$$

where $\sigma_s^*(\eta) = \sigma_s^*(\omega(\eta))$, $R_{\mathrm{I}}(\eta)$ – dimensionless normal stress along the contour of the notch.

At the tip of the rounded notch ($\eta = 0$) we have:

$$\sigma_s^*(0) = (2\pi\rho)^{-\lambda_I}\,\tilde{K}_I^V\,R_I,$$

where $R_I = R_I(0)$ – the stress rounding factor of the V-notch on the stress at its vertex (Savruk & Kazberuk 2017, Pilkey W. & Pilkey D. 2008), which depends on the vertex angle of the notch, its shape in a certain neighborhood of the vertex, and also on the elastic constants of the orthotropic material. The relation for the semi-infinite rounded V-notch is exact, and for restricted bodies it is asymptotic and is more accurate the smaller is the dimensionless radius of notch tip curvature. It can be used both to estimate the normal stress at the vertex of the notch, and to find the stress intensity factor at the tip of the acute V-notch from the ratio

$$\tilde{K}_I^V = \frac{1}{R_I}\lim_{\rho\to 0}[(2\pi\rho)^{\lambda_I}\sigma_s^*(0)].$$

To calculate the stress intensity factor $\tilde{K}_I^V$ at the tip of an acute V-notch, it is necessary to know both the stress rounding factor R_I and the stress singularity exponents λ_I.

For quasi-orthotropic material (A degenerate orthotropic material whose shear modulus G is related to other material dependency characteristics $G = E_1/[2(\sqrt{E_1/E_2} + \nu_{12})]$ (Savruk & Kazberuk 2017) we have

$$\lambda_I = 1,247\cos\beta_1(\alpha) - 1,312\cos^2\beta_1(\alpha) + 0,8532\cos^3\beta_1(\alpha) - 0,2882\cos^4\beta_1(\alpha),$$

where $\beta_1(\alpha) = \pi + \mathrm{arctg}(\gamma\mathrm{tg}\alpha)$.

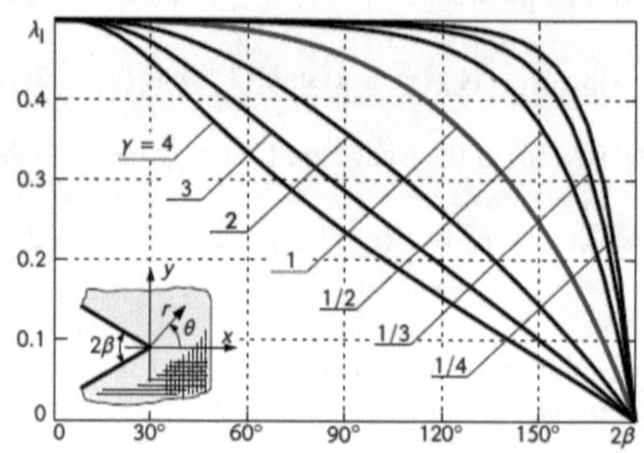

Figure 2. V-notch singularity order λ_I as a function of notch opening angle 2β for quasi-orthotropic material.

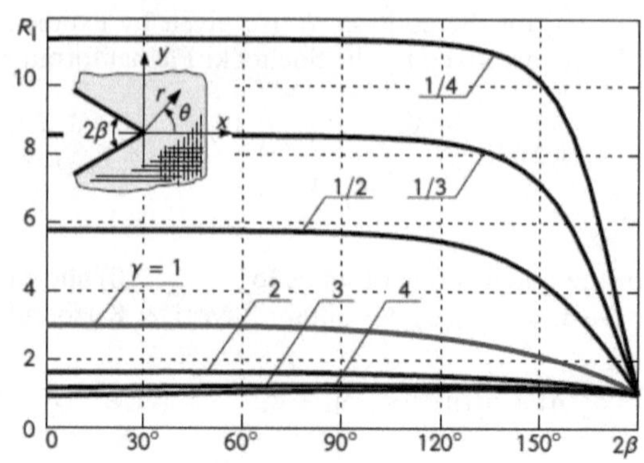

Figure 3. Stress rounding factors R_I at the rounded notch vertex as a function of notch opening angle 2β for quasi-orthotropic material.

The numerical dependencies of the stress singularity exponents λ_I and the stress rounding factor R_I of the V-notch on the vertex angle 2β are shown in Figures 2 and 3 for quasi-orthotropic material.

In a quasi-orthotropic wedge, the values of the relevant parameters depend significantly on the orientation of the wedge bisector of the orthotropic axes.

REFERENCES

Khoma, M.S., Korniy, S.A., Vynar, V.A., Datsko, B.M., Maksishko, Yu.Ya., Dykha, O.V., Bukliv, R.L. 2022. Influence of hydrogen sulfide on the carbon-dioxide corrosion and the mechanical characteristics of high-strength pipe steel. *Mater. Sci.* 57: 805–812.

Lavrys, S.M., Pohrelyuk, I.M., Lukyanenko, A.G. 2023. Fatigue limit of two-phase titanium alloy after surface deformation-diffusion treatment. *JOM*. 75: 1251–1260.

Savruk, M. P., Kazberuk, A. 2017. *Stress Concentration at Notches*. – Cham: Springer.

Kazberuk, A., Savruk, M.P., Chornen'kyi, A.B. 2016. Stress concentration near an elliptic hole or a parabolic notch in a quasiorthotropicplane. *Mater. Sci.* 52: 295–304.

Ioakimidis, N. I., Theocaris, P. S. 1977. The problem of the simple smooth crack in an infinite anisotropic elastic medium. *Int. J. Solids Struct.* –13(4): P. 269–278.

Pilkey, W.D., Pilkey, D.F. 2008. *Peterson's Stress Concentration Factors*, 3 ed. New York: Wiley.

Welding and Related Technologies – Krivtsun et al. (Eds)

Mechanical properties of welded joints of medium carbon alloyed steels heated to temperatures from 250 to 1200 °C

V.D. Pozniakov, O.A. Gaivoronskiy, A.V. Zavdoveev, O.V. Korenev & O.V. Makhnenko
Paton Electric Welding Institute of NAS of Ukraine, Kiev, Ukraine

ABSTRACT: During welding two characteristic areas are formed - hardening and tempering. The depth of these areas reaches 2.5 mm and 8 mm, respectively. Depending on the steel grade, the depth and value of hardness of these areas change. The transition from the first to the second occurs in the area of incomplete recrystallization, which in structure is a mixture of martensite, ferrite and carbides. The total width of the de-hardened area of the heat-affected zone (HAZ), consisting of areas of incomplete recrystallization and tempering, for welded joints with a thickness of 6 - 20 mm and depends on the linear energy of welding and can vary in the range of 3 … 8 mm. The work provides for the first time structured information on the features of the thermal welding cycle and its influence on the mechanical properties in the HAZ metal in the area heated to temperatures from 250 to 1200°C.

1 INTRODUCTION

Steel grade "71" refers to heat-strengthened medium-carbon alloy steels of high hardness, HB over 5000 MPa, and strength, σ_{in} more than 1400 MPa. Like most similar steels, such as ARMSTAL 500 (Poland), Armox 500 (Sweden), Mars 500 (France), Miilux Protection 500 (Finland) and others acquire high hardness and strength due to alloying with manganese, silicon, chromium, molybdenum, nickel, microalloying with boron, titanium, aluminum, vanadium, etc., as well as due to heat treatment, which consists of hardening the steel and its low-temperature tempering at a temperature not exceeding 250 °C (Özdemir 2020., Konat et al. 2017, Gaivoronsky et al. 2023, Oskwarek et al. 2006, Cabrilo et al. 2016, Kuzmikova 2013, Arsić et al. 2014).

The specified steels are classified as armor steels. They are capable of withstanding significant dynamic loads and have good wear resistance. Such steels can be used both for the manufacture of military products subject to impact loads and for the production of civilian machines and mechanisms, such as quarry excavator buckets, heavy-duty vehicle bodies, etc. Most of these products are welded and are overwhelmingly welded using arc processes. This process involves local heating of rolled steel to temperatures exceeding the tempering temperature of steel and reaching 1200 ° C. The dimensions and distance from the weld axis of the sections that are heated to the specified temperatures in the welded joint depend on many factors. The main ones are the welding mode, metal thickness, and joint type. The listed factors also affect the cooling rate of the HAZ metal of welded joints. In turn, the cooling rate determines what structure will be formed in the HAZ metal heated above the Ac3 temperature (Zavdoveev et al. 2021).

Usually, studies on the influence of welding modes on the mechanical properties of various sections of the HAZ of welded joints involve assessing their rigidity and impact toughness, and in some cases, the ability to resist brittle fracture. Samples for such studies can be made directly from welded joints. Thus, the purpose of this work was to determine the influence of arc welding modes on the heating and cooling conditions of metal in various sections of the HAZ, reproduce these conditions during heat treatment of the studied steel blanks, and

DOI: 10.1201/9781003518518-41

establish how they affect the mechanical properties of the metal in local sections. joints that are heated to a temperature of 250 °C and higher.

2 METHOD

The studies were conducted on high-strength armored steels with a yield strength of more than 1200 MPa of domestic and foreign production of steel grades 71, ARMOR 500T, PROTECTION 500 and RAMOR 500, respectively, the chemical composition of which is given in Table 1. High-alloy wires of solid cross-section with a diameter of 1.2 mm, widely used in domestic and European practice of welding production of LBT, were selected as welding materials.

Table 1. Chemical composition of high-strength and high-rigidity armor steels 500HV.

		Mass fraction of elements, %									
No.	Steel, brand	C	Si	Mn	Cr	Ni	Mo	V	S	P	B
1	71*	0.31	1.16	0.74	1.66	2.26	0.30	0.202	0,010	0,016	—
2	Protection500*	0.28	0.49	0.96	0.58	0.37	0.25	0.002	0,011	0,016	0.002
3	ARMOX500T	0.23	0.25	0.82	0.50	0.90	0.30	0,030	0.004	0,016	0.003
4	RAMOR500*	0.21	0.45	0.92	0.58	0.38	0.2	0.01	0.005	0.023	0.004

To ensure a constant welding speed, a specialized trolley with a drive was used in the experiments, ensuring a stable torch travel speed in the typical welding speed range of 10...25 m/h. Welding was performed by the mechanized method in a protective gas environment (82% Ar + 18% CO_2) using KHORDA 307 Ti (Sv-08Kh20N9G7T) wire with a diameter of 1.2 mm, with preliminary heating to a temperature of 100 °C and without heating at welding currents from 180A to 300A.

The location of the tempering section was calculated based on preliminary studies of the microstructure on specially prepared sections and measurements of the microhardness of the metal in the heat-affected zone with a step of no more than 1 mm. Mechanical characteristics of the tempering section (hardness and impact toughness) were determined at a distance of up to 4-5 mm from the fusion line.

3 RESULTS AND DISCUSSION

As can be seen, the hardness of the welded joint of grade 71 steel in the HAZ metal in the tempering area remains at the level of the hardness of the base metal 507HV when welding in the standard mode, both without heating and with heating. Welding in the forced mode, under the same conditions, causes a weakening of about 8 ... 10% in grade 71 steel. Such behavior of grade 71 steel (Table 2) is explained by the increased degree of alloying elements that sufficiently restrain the tempering of the metal in the potentially vulnerable HAZ area under the action of the welding thermal cycle (WTC).

ARMOX 500 T steel indicate that it is too vulnerable to the destructive action of WTC. It was found that in the HAZ, neutralization of about 12% and 36%, respectively, is observed during welding in standard and forced modes, both without and with heating. is associated with the fact that the creation of high strength and rigidity in the production of rolled products is carried out mainly due to heat treatment and due to more economical alloying.

As we can see, the impact toughness values of the base metal of armor steels of the PROTECTION 500; ARMOX 500 T and 71 grades naturally decrease with decreasing test temperature, but do not cross the minimum permissible limit according to the NTD requirements of $KCU_{-40\ °C}$ <35 J/cm2. The highest impact toughness values are characteristic of steel grade 71, then in the rating is the PROTECTION 500 grade, the third is the ARMOX 500 T grade, which is in good agreement with their alloying and the original structure after heat treatment.

Results indicate that at the test temperature of "+20 °C" and "-40 °C" the pulse current welding modes (standard and forced) have virtually no effect on the impact toughness of the HAZ metal in the tempering section within the studied grades of armor steels. The general pattern of the arrangement of PROTECTION 500, ARMOX steels is preserved 500T and 71 in the rating depending on the degree of alloying.

Table 2. Influence of welding modes on the hardness of the HAZ metal in the tempering area.

Item No.	Mark became and	Welding mode (IDZ)	Welding conditions	Hardness, HB
1	71	standard	Without heating.	507
			Heating 100 ° C	507
	71	forced	Without heating.	450
			Heating 100 ° C	460
	71 base metals	delivery status		507
2	ARMOX 500 T	standard	Without heating.	430
			Heating 100 ° C	364
	ARMOX 500 T	forced	Without heating.	307
			Heating100 About S	337
	ARMOX 500 T	delivery status		493
3	PROTECTION 500	standard	Without heating.	431
			Heating 100 ° C	431
	PROTECTION 500	forced	Without heating.	420
			Heating 100 ° C	420
	PROTECTION 500	delivery status		478

When determining the effect of preheating temperature on impact toughness KCU $_{-40\ °C}$ in the tempering section of the HAZ metal during welding in standard and forced modes, its insignificant effectiveness was established for the studied steels of the PROTECTION 500, ARMOX 500 T and 71 grades.

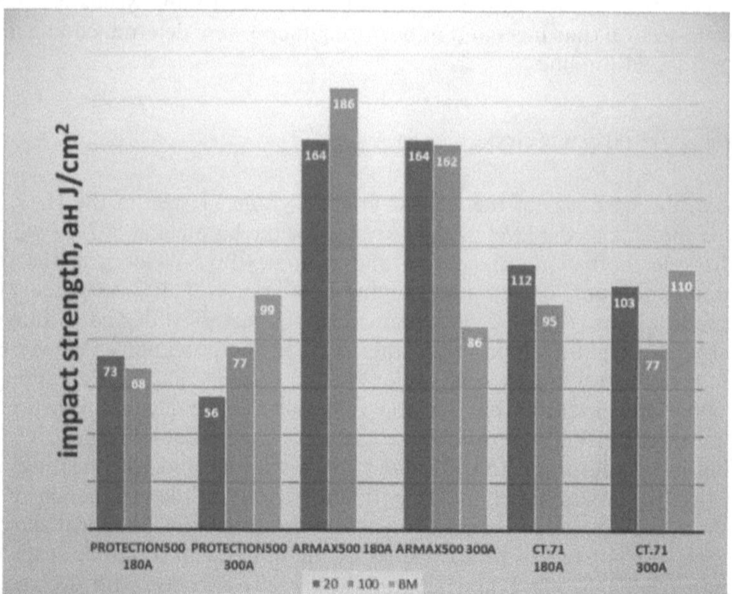

Figure 1. Impact toughness indicators of welded joints of armor steels under different welding conditions, tested temperature is -40°C.

In the area of hardening of the metal of the HAZ of welded joints of armor steels, the impact toughness indicators are significantly lower. For example, for welded joints of armor steel grade ARMOX 500 T during IDS in the standard mode, the impact toughness indicators during sample examinations at a temperature of +20°C are more than 1.5 times lower than the indicators for the tempering zone, and on average amount to 94 J/cm2 (Figure 1). When welding in the forced mode, they increase to values of 115 J/cm2 · At the same time, the resistance to destruction under impact in the hardening zone is close to the base metal of the armor steel, which is 97 J/ cm^2.

4 CONCLUSION REMARK

Thus, it has been established that the metal of the heat-affected zone in the areas of hardening and tempering of welded joints of armor steels of foreign manufacture with a hardness of 500 HB during pulsed-arc welding has sufficient resistance to destruction under impact loading compared to the base metal. At the same time, in the tempering area, the impact toughness index is more than 1.5 times higher than in the base metal.

REFERENCES

Arsić, D., Nikolic, R., Lazic, V., Hadzima, B., Aleksandrovic, S. and Djordjevic, M., 2014. Weldability estimates of some high strength steels. *XLII. International Conference 2014 VERSION Tatranská Lomnica*, pp.12–14.

Cabrilo, A. and Geric, K., 2016. Weldability of High Hardness armor steel. *Advanced Materials Research, 1138*, pp.79–84.

Gaivoronsky, O., Podnyakov, V., Zavdovyev, A., Klapatyuk, A., Denisenko, A., 2023. Prevention of bad jumps in reinforcement rack linen. Paton Welding Journal, 5, pp. 3–10.

Konat, Ł., Białobrzeska, B. and Białek, P., 2017. Effect of welding process on microstructural and mechanical characteristics of Hardox 600 steel. *Metals*, *7*(9), p.349.

Kuzmikova, L., 2013. Intellectual tendency towards high hardness armor steel, *Wollongong Australia* pp.17–234.

Oskwarek, M., 2006, May. Structural features and susceptibility to cracking of welded joints of Hardox 400 and Hardox 500 steels. In *Conference Proceedings, IV Scientific Conference of Students, Wroclaw University of Technology Publishing House, Wrocław*.

Özdemir, T., 2020. Mechanical & microstructural analysys of armor steel welded joints. *International Journal of Engineering Research and Development*, *12*(1), pp.166–175.

Zavdoveev, A., Poznyakov, V., Baudin, T., Rogante, M., Kim, H.S., Heaton, M., Demchenko, Y., Zhukov, V. And Skoryk, M., 2021. Effect of nutritional values on the processing properties and microstructure of HSLA rod processed by different technologies. *Materials today communications*, 28, pp.102598.

Author Index

Kolibri 360 GmbH
Ferdinand-Jühlke-Straße 7
99095 Erfurt, Deutschland
produktsicherheit@kolibri360.de